Optical Technologies in the Humanities

Springer
Berlin
Heidelberg
New York
Barcelona
Budapest
Hong Kong
London
Milan
Paris
Santa Clara
Singapore
Tokyo

Series of the International Society
on Optics Within Life Sciences

OWLS

Series Editor:
Gert von Bally *Laboratory of Biophysics, Institute of Experimental Audiology, University of Münster, Robert-Koch-Str. 45, D-48129 Münster, Germany*

Volume IV

**International Society
for Historical Geography of the Ancient World
EUROCARE**

OWLS IV
**International Society on
Optics Within Life Sciences**

International Commission for Optics

University of Münster

D. Dirksen · G. von Bally (Eds.)

Optical Technologies in the Humanities

Selected Contributions
to the International Conference
on New Technologies in the Humanities
and
Fourth International Conference
on Optics Within Life Sciences OWLS IV
Münster, Germany, 9–13 July 1996

With 169 Figures and 7 Tables

Springer

Editors:

Dieter Dirksen
Gert von Bally

Laboratory of Biophysics
Institute of Experimental Audiology
University of Münster
Robert-Koch-Str. 45
D-48129 Münster, Germany

Die Deutsche Bibliothek - CIP-Einheitsaufnahme

Optical technologies in the humanities : selected contributions to the International Conference on New Technologies in the Humanities and Fourth International Conference on Optics Within Life Sciences OWLS IV, Münster, Germany, 9 - 13 July 1996 ; with 7 tables / [International Society on Optics within Life Sciences]. Ed. by D. Dirksen and G. von Bally. - Berlin ; Heidelberg ; New York ; Barcelona ; Budapest ; Hong Kong ; London ; Milan ; Paris ; Santa Clara ; Singapore ; Tokyo : Springer, 1997
(Series of the International Society on Optics within Life Sciences ; Vol. 4)
ISBN-13: 978-3-642-64595-2

Library of Congress Cataloging-in-Publication Data applied for

ISBN-13: 978-3-642-64595-2 e-ISBN-13: 978-3-642-60872-8
DOI: 10.1007/978-3-642-60872-8

Softcover reprint of the hardcover 1st edition 1997

Typesetting: Camera ready by authors/editors
Printed on acid-free paper
SPIN 10534344 57/3144 - 5 4 3 2 1 0

Preface

New high-tech developments show increasing applicability not only in classical technological fields but also in the humanities. Here research is confronted with questions whose solutions demand the application of new natural scientific methods and instrument engineering. Thus, a dialog between cultural scientists and engineers is necessary and should be strengthened beyond the actual individual basis. The interdisciplinary "International Conference on New Technologies in the Humanities" — the first of its kind — which gives the background for this book, demonstrated that such a dialog can be extended to an international level profitable for both sides. It showed the range of available methods and their potential for cultural aspects, as well as challenging extentions of the usual tasks for applied scientists and engineers.

The International Society on Optics Within Life Sciences (OWLS) was founded in 1990 as a non-profit scientific organization based on individual membership. Its objective is to forward interdisciplinary information and communication of specialists in optics as well as in medicine, biology, environment and cultural heritage.

To serve this purpose a series of *International Conferences on Optics Within Life Sciences (OWLS I-IV)* has been established. The latest in this sequence (OWLS IV) was held in Münster, Germany, 9 - 13 July 1996, jointly organized with the above mentioned "International Conference on New Technologies in the Humanities" and dedicated to the key topic "Optical Technologies in the Humanities".

Cosponsoring Organizations were:

- International Society on Optics Within Life Sciences (OWLS)
- International Commission for Optics (ICO)
- International Society for Historical Geography of the Ancient World (Ernst-Kirsten Society)
- EUROCARE
- German Oriental Society
- Ministry of Education, Research, Science and Technology of the Federal Republic of Germany
- University of Münster
- Research Center Jülich

This volume comprises selected contributions to both conferences. 40 papers submitted by authors from 14 countries cover international research activities in the following areas:

- Archaeological Research and New Technologies

- Holography and Other Interferometrical Techniques
- Other Analytical Techniques (Material Analysis, Dating, Localization, Biomolecular Methods, etc.)
- Lasercleaning
- Pattern Recognition including Digital Evaluation
- Unconventional Microscopy
- Spectroscopical Techniques (Fluorescence, Luminescence, Reflection, Absorption)
- Profilometry

The editors wish to thank Ms. J. Mand for her assistance in preparing the manuscript for this volume. We also gratefully acknowledge the financial support by Coherent GmbH and Polytec GmbH.

D. Dirksen, *Münster, Germany*
G. von Bally, *Münster, Germany*

Other volumes within this series:

OWLS I *Optics in Medicine, Biology and Environmental Research*, G. von Bally, S. Khanna (editors), Elsevier, 1993

OWLS II *Optics for Protection of Man and Environment against Natural and Technological Disasters* G. von Bally, H. I. Bjelkhagen (editors), Elsevier, 1993

OWLS III *Optical Methods in Biomedical and Environmental Sciences* H. Ohzu, S. Komatsu (editors), Elsevier, 1994

Contents

Pattern Recognition, Including Digital Evaluation

Unconventional Microscopy

Spectroscopical Techniques (Fluorescence, Luminescence, Reflection, Absorption, etc.)

Archaeological Research and New Technologies

New Technologies in Cypriote Archaeology: A Current Research Program on Ancient Ceramic Technology

Vassos Karageorghis

Archaeological Research Unit, Univ. of Cyprus, P.O.Box 537, CY1678 Nicosia, Cyprus

Nota Kourou

Department of Archaeology, University of Athens, 157 84 Athens, Greece

Eleni Aloupi

THETIS Hellas Ltd, 41 M.Moussourou St., 116 36 Athens, Greece, E-mail: thetis@netor.gr

Abstract. The paper presents preliminary results of a close collaboration of scientists and archaeologists in the framework of a research program that refers to a large scale systematic study of ancient ceramic technology in Cyprus from the Late Neolithic to the Classical period, making use of modern analytical techniques, computer-aided data handling and evaluation software.

1 Introduction

The aim of this paper is to demonstrate how already known scientific techniques are being applied in Cypriote archaeology today and how "traditional" archaeologists are collaborating with scientists in resolving specific problems which traditional methods have not been able to resolve.

New technology was applied in Cypriote archaeology already in the 1950's, when carbon-14 analysis helped to date correctly the Cypriote Neolithic period to the 7th and 8th millennia BC whereas it was hitherto dated by traditional methods to the 4th millennium BC. Apart from the C-14 technique which was introduced by Libby in the late 1940's, there are now other methods for dating (e.g. thermoluminescence, potassium-argon dating, fission tracks, EPR dating etc.) which have been used extensively in archaeological research to the extent that they constitute the scientific core of a quantitative discipline which in some quarters has been called *Archaeometry*. On the other hand, after the 1960's we have witnessed the emergence of a new systematic multidisciplinary approach to archaeological issues. This approach is to be distinguished from archaeometry in the sense that the members of the multidisciplinary team (i.e. the chemist, physicist, geologist, oceanographer, botanist, palinologist, geologist etc.) come together for a specific project. This multidisciplinary approach has been adopted mainly in cases where there are difficulties in the interpretation of the archaeological material with traditional methods, mainly in prehistoric archaeology. In contrast, the traditional archaeologist deals with such a wealth of information that he is much less motivated to seek the collaboration of natural scientists. The situation presents an analogy with a criminal investigation where the detective in charge will not seek the help of the forensic scientist if the criminal he has arrested was standing next to the victim, with blood on his hands.

The usefulness of a multidisciplinary approach emerges when it is preceded by a thorough archaeological study of a large body of data, which allows the formulation of clear-cut questions and helps to give appropriate answers. It is in this spirit that scientists from different disciplines were brought together in order to investigate the ancient ceramic technology in Cyprus from the Neolithic to the Hellenistic periods.

2 Multidisciplinary Study of Cypriote Ceramic Technology

The geographical situation of Cyprus at the cross-roads of three continents has exposed the island to contacts with and successive occupation by a wide range of cultures which have left their mark on local artefacts (Myceneans and Cretans, Phoenicians, Egyptians and others). Ceramic artefacts in particular provide excellent material against which these interactions can be studied since they contain multidimensional information with respect to the shape, the style of decoration (incised, painted, plastic), the fabric, the raw materials used, the manufacturing techniques etc. It is now widely recognised that the investigation of ancient ceramic technology, which was usually based on the analysis of the ceramic body in the past, can be complemented through the analysis of the pigments used for the surface decoration [1].

A preliminary study of a small number of Cypro-Archaic polychrome terracotta figurines from the Louvre Collection (figure 1) by using the PIXE (Proton Induced X-ray Emission) non-destructive technique [2], revealed the use of raw materials rich in manganese oxides for the black colour (i.e. umbers which are abundant in Cyprus). This confirmed the application of the Mn-black decoration technique in the Cypro-archaic I and II periods, at a time when it had been abandoned in Greece [3] and totally replaced by the iron reduction technique. The analysis also revealed [2] the presence of cinnabar (mercury sulphide, HgS) in the red pigment of one of the figurines representing a horse and rider (the one on the right in figure 1). Interestingly, the blue-green pigment on the same figurine was identified as a zinc-based material, probably zinc carbonate (smisthonite). These observations provide a contrast with the most frequent use of ochre for the red and green earth (celadonite) for the green, detected in the rest of the figurines which have been analysed (figures 2a,b,....). The latter two minerals (i.e. iron oxides and green earth) are abundant in Cyprus, while cinnabar and smisthonite are not known to be present on the island. A plausible explanation was that the use of such pigments, probably imported from Anatolia or even Spain, characterises the production of a distinct ceramic workshop. The consolidation and interpretation of this suggestive evidence can only be achieved through a large scale systematic study which has been undertaken during this project.

The ceramics in the Cyprus Museum in Nicosia provide a complete and comprehensive archaeological collection for the study which spans more than 40 centuries from Neolithic to Hellenistic times. Due to the nature and wealth of the material, the first step of the project consisted of an in situ survey using non-destructive analytical techniques, in conjunction with digital recording of visual information (digital camera, 3-D image recording system). Recent developments in X-ray detection have led to high resolution compact, portable and low cost instruments operating at room temperatures. The use of this new technology offers new possibilities for a conventional analytical technique such as X-ray Fluorescence (XRF) and has made this survey possible.

Fig. 1. Cypro-Archaic polychrome terracotta figurines of the Louvre Collection.

An early, practically on line result of the survey was that no further evidence for the use of cinnabar in red pigment was found in the Cyprus Museum collection of all figurines on which the red paint was still preserved. The X-ray analytical data showed that the red pigment was an iron rich material obviously obtained by the use of ochre. A collection of 43 figurines was examined in less than two hours. This illustrates the power of new technology in a case where the archaeological question is very specific. In view of these results it is probably safe to conclude that the presence of mercury sulphide (i.e. cinnabar) in the single figurine in the Louvre Museum must be attributed to post excavation retouching.

The XRF analysis of 75 ceramic artefacts revealed the alternate use of the two different techniques for the production of black colour (figures 3a,b) which are known as manganese black and iron reduction techniques. This *interplay* between Mn - rich and Fe - based black indicates the use of very different

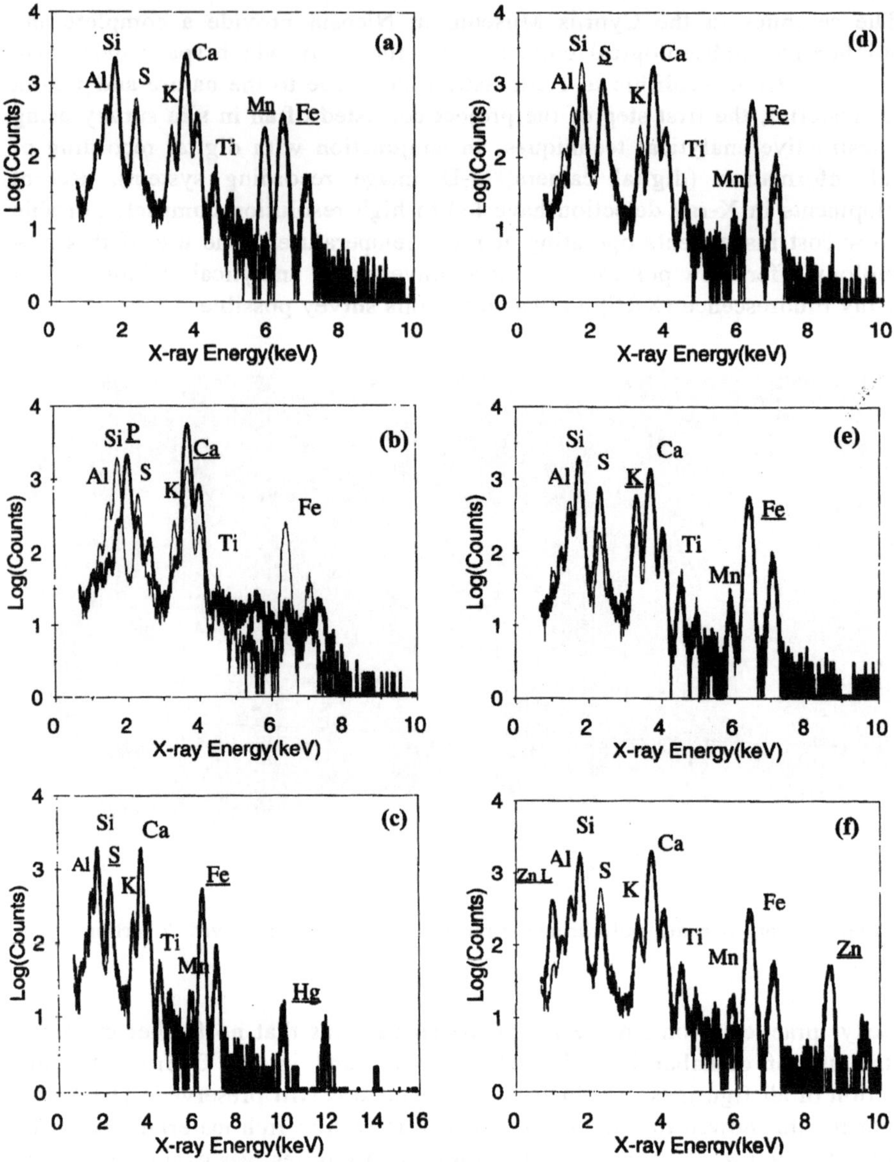

Fig. 2. Typical X-ray spectra obtained by PIXE analysis of Cypriote terracotta figurines from the Louvre Collection. The thick line corresponds to the painted areas and the thin line to the body. Relevant elements detected are underlined: (a) Mn-black, umber, (b) white, Ca and P (pulverised bone?), (c) Hg-rich red, cinnabar, (d) Fe-rich red, ochre, (e) green, green earth, (f) bluish-green, Zn based pigment (smisthonite?)

firing which consequently point to different technological traditions. The latter, seen in the context of the different ethnic origin of the various potters (native Cypriots, Cretans, Myceneans, Syrians, Phoenicians) during several periods, may suggest either the introduction of new production techniques or the resistance of local tradition to external influence.

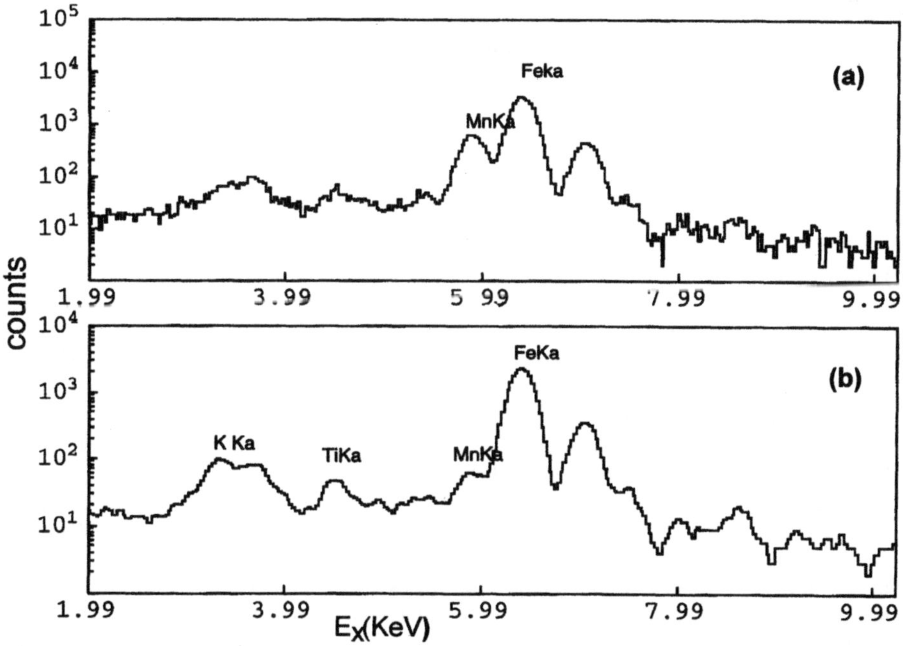

Fig. 3. Typical XRF spectra of the dark paint decoration of two ceramic objects that follow the White Slip I and II styles, respectively. (a) Mn-rich black pigment, (b) Fe-based black produced by the iron reduction technique.

Following this initial survey which shed light on the surface decoration techniques and materials used, certain categories of the Cyprus Museum collection (i.e. White slip pottery, local imitations of Mycenean, Protowhite painted, Protobichrome, early Geometric, Cypro-Archaic and Cypro-Classical) will be studied in more detail by applying destructive techniques, mainly Scanning Electron Microscopy with Energy Dispersive Microanalysis equipment (SEM/EDX), petrographic analysis, X-Ray Diffraction (XRD) and Neutron Activation Analysis (NAA) on ceramic sherds specifically sampled to match the characteristics of the most interesting exhibits. Special attention was given to the selection of ceramic samples in collaboration with the archaeologists. The above methodology, based on the application of several analytical techniques, will be supplemented by reproduction experiments under laboratory conditions.

Throughout the project, the management and processing of the large body of data consisting of analytical, visual and archaeological information will make use of digital techniques in conjunction with the low cost data storage capacity of personal computers and compact discs. Special emphasis has been given to the synthesis of the research team in order to complement one another. For this reason, a chemist specialising in archaeological ceramics, two physicists, a geologist, a petrologist, and a computer scientist were brought together with two archaeologists [5] as well as a ceramic artisan who, by the end of the study, will replicate some representative artefacts by following the ancient manufacturing techniques which will be elucidated through this project.

References

1 E. Aloupi, *Nature and Micromorphology of Paint Layers on Ancient Ceramics. A New Approach to the Study of Ancient Ceramic Technology* (Ioannina,1994; ISBN 960-90007-0-3)

2 E. Aloupi and D. McArthur in *The Coroplastic Art of Ancient Cyprus*, Vol IV (A.G. Leventis Foundation, University of Cyprus, Nicosia, 1995)

3 Jones, R.E., *Greek and Cypriot pottery*, 798-805. The British School at Athens, Fitch Laboratory Occasional Paper 1, Athens, (1986)

4 Svardh, A., Analysis of pigments in painted terracottas in G. Ikosi, "Kythrea Temenos. Unpublished Material from the Swedish Cyprus Expedition," *Medelhavsmuseet Bull*. **28**, 42-49. 1993

5 E. Aloupi, A. Karydas, P. Kokkinias, D. Loukas, T. Paradellis, A. Lekka, and V. Karageorghis, in *Proceedings of the 3rd Symposium on Archaeometry of the GSA*, (The Greek Society for Archaeometry, Athens, forthcoming)

The Role of Science in Archaeological Regional Surface Artefact Survey

John Bintliff,
Department of Archaeology, Durham University,
Durham, U.K.

My purpose in this paper is to address the question - how far today is the scientific field archaeologist, utilizing essentially surface remains, able to approach the complex behaviour of past peoples in space which the imaginative reconstructions of Classicizing painters such as Lorraine and Poussin present to us? (Fig. 1) I shall be using mainly Greek examples, to answer that challenge.

Fig. 1: The Burial of Phocion, by Poussin.

The Development of Extensive Survey

By the late Nineteenth Century, skilled topographers such as Lolling [1] (1876-7) were travelling exhaustively along the byroads and paths of South European countries, noting every standing ruin of every age. Maps such as *the Atlas de la Grèce* (1852) record both modern settlements, ancient ruins and deserted villages of post-Medieval date. But essentially these researches record the more obvious, larger archaeological sites where walls or mounds of rubble and tile were visible even at some distance. Only gradually was the idea being put into practice where the fieldworker wandered around small areas of landscape looking for more vestigial traces of settlement - notably those where only surface pottery betrayed past settlement and activity. An excellent example was being provided by German scholarship, where the importation of the flourishing tradition of *Landschaftsarchäologie* with its concern for the shifts of settlement within definable *Siedlungskammer* produced pioneering papers on the long-term utilisation of small landscapes, which were sadly to be neglected till very recently (e.g. Lehmann, [2]). With the retreat of German scholarship in the postwar era, American influences and examples came to the fore. A new standard for regional field survey was set by the massive University of Minnesota Expedition to the province of Messenia [3]. A very large region of the south-west Peloponnese was extensively travelled over, whilst both ceramic and standing monument surveys were conducted. The focus was especially on the Bronze Age settlements, but a striking novelty lay in the great range and depth of the interdisciplinary scientific contributions to the final project volume, chiefly in the realms of physical and human geography. Notably at around the same time, a similarly exemplary, largescale regional survey was setting a new standard in extensive survey for Italy - the British survey of South Etruria in Italy [4].

The Intensive Survey Revolution

Methodological debate in the United States, especially associated with the theoretically-orientated 'New Archaeology' movement, led during the 1960s to a demand for far more rigorous modes of landscape study, in order to allow evaluation of the total number and nature of activity and settlement traces of each period across defined areas of the landscape. This meant the use of lines of fieldwalkers some 5 to 50 metres apart, walking large continuous areas of countryside and trying to record all surface traces of past behaviour - especially bits of broken pottery. As in the States, it was quickly shown that these 'intensive' surveys found an incomparably higher number of sites per square kilometre than previous approaches. At first, generally following American schemes, regions were sampled using a scatter of survey blocks, or strips separated by unsurveyed land; total survey was carried out in the sample units, then their results provisionally extrapolated to the remaining countryside on the assumption of their representativity (e.g. on the Melos Survey, one-fifth of the island was fieldwalked in parallel 1-km wide strips each 5 kilometres apart) (Fig. 2)[5].

Whilst this approach dominated projects carried out in the 1970s and early 1980s, it was gradually being realized that the use of such sampling lacked empirical justification and was primarily being done to allow the study of large areas within a few seasons of research. Settlement patterns known historically were seen to be irregular rather than regular, whilst sampling systems worked much better for landscapes with innumerable, well-dispersed settlements than for landscapes where most people lived in a few nucleated sites. Moreover, further theoretical and empirical experiments in the States had shown that in long-occupied landscapes the surface archaeology could reveal not only very small occupation sites but also places

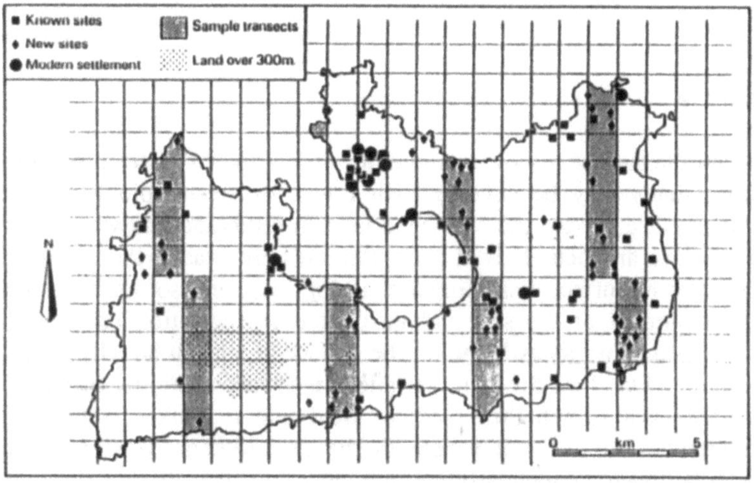

Fig. 2: The 20% sample scheme adopted for the Melos Survey (from [5] fig.2.1)

of non-residential activity ('offsite activity'). Clearly the surface of the countryside was much more complex and unpredictable than had been imagined, and the best response was to slow down the speed of survey teams, and attempt to fieldwalk as close to 100% of the landsurface as possible. Moreover, as the concept was spreading of the entire landsurface as a human 'artefact', blurring the distinction between 'site-settlement' and a varying intensity of human behavioural traces beyond settlements (offsite scatters), a new methodology was required to record the entire surface. On my own project, the British Academy Boeotia Project (co-directed by Anthony Snodgrass of Cambridge University), in 1980 we adopted hand-held 'clickers' or manual counting devices, so that each fieldwalker could count every visible artefact in their path as they marched across the gridded landscape. Areas where the surface finds were very high or very fresh, or accompanied by building or other dense-activity debris, were revisited as potential 'sites' - and increasingly we were able to differentiate villages, farms, and rural cemeteries from each other [6] [7].

Gridding the entire landscape so that all of its surface archaeology was mapped was an approach many projects also adopted for the individual site found within the fieldwalking grids, so that the patterning of surface artefacts across the landscape could be followed into the microscale, complex variable patterning around and across settlement sites within those landscapes. By the end of the 1980s the advent of inexpensive hand-held computers and of Geographical Information Systems (GIS) offered ideal opportunities for adapting such intensive recording systems to the investigative power of elaborate spatial analytical software programmes. The search for patterns in surface landscape data and their matching to every kind of spatial variable now became a relatively simple task. I shall return later to the potential of GIS in the conclusion to this paper.

Correction Factors

Relying essentially on the surface discovery and plotting of artefacts raises problems of surface visibility; fields may be covered with crops, or freshly-ploughed, or fieldwalking may take place beyond the cultivated land in scrubland and woodland. As is well-known, surface artefacts derived from past sites are usually the result of regular ploughing of the subsurface disturbing archaeological levels, so that both the current and recent treatment of the soil and its degree of vegetation cover will seriously affect the density of artefacts found on the surface of ancient activity areas. A simple but effective way to allow for these variations is to take 'visibility readings' of the landsurface for each stretch ('transect') or field that is walked, reflecting numerically the percentage of the soil visible to the fieldwalkers. The count of surface artefacts can be given subsequently not only as a 'raw' density, but through multiplication using the visibility factor a more realistic corrected density can be mapped (Figs. 3-4; from [8]).

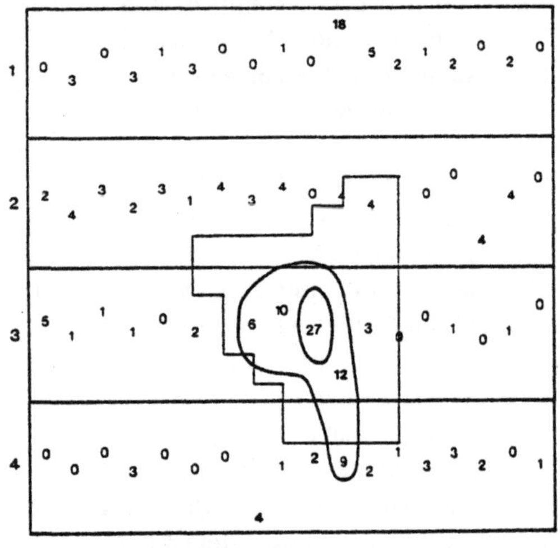

Fig. 3: Fieldwalking transects on the Hvar (Croatia) Survey, raw counts of surface pottery along individual north-south strips each 45m long and 1m wide. A clear anomaly in the centre of this sector suggests a settlement site (from [8] fig.9.7a).

The instability of the soil over time should also be investigated. Especially in Southern Europe, processes of erosion and deposition are widespread, giving different sectors of the same landscape a very variable soil development from ancient and prehistoric times to today. In some areas erosion may result in high site visibility, with very little left beneath the soil; in others more stable soil development may limit disturbance of buried deposits to thin scatters on the surface; in yet other areas, colluvial or alluvial deposition may partially or completely seal ancient landsurfaces and prevent sites from being made visible by ploughing. The services of a geomorphologist and soil scientist are seen today as essential to map in advance the

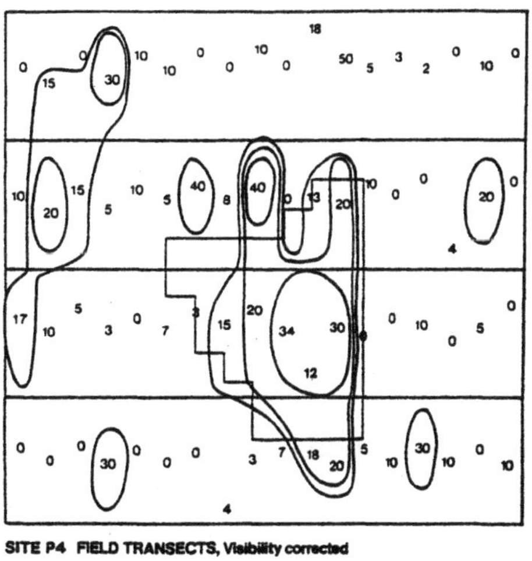

Fig. 4: The same sector as Figure 3, but with surface ceramic counts corrected for variable surface visibility. The settlement site is more clearly defined and more extensive, but there is another minor anomaly to be studied. Detailed site survey confirmed that the main anomaly was a large Roman villa (from [8] fig.9.7b)

distribution of such 'taphonomic' types, so as to avoid misleading inferences on the apparent density of activity as it appears today being a simple reflection of past activities (cf. [9]).

A further observation follows from what has just been said about erosion, deposition and cultivation disturbance to archaeological landsurfaces; over time one may expect that settlements and other activity areas suffer progressive attack by both natural and human weathering processes. Other things being equal, this means that a typical farmsite of High Medieval times will be better preserved on the surface and in the subsoil than one of Classical Greek times, whilst the same small farm of Early Bronze Age date will very likely be found in an extremely vestigial or minimal condition today. Figures 5-7 illustrate this using a small ancient farmsite in the Valley of the Muses in Boeotia: the first figure shows the total density of surface pottery per square metre, the second the number of securely-dated Classical sherds, the final that of Late Bronze Age pottery. The reasonable number of Classical pieces

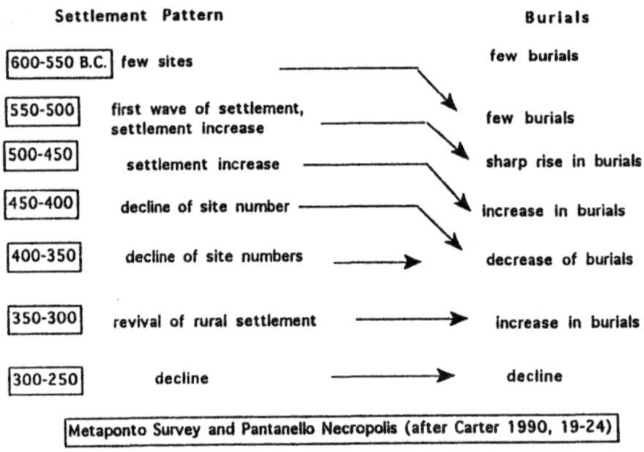

Fig. 8: Correlation between the rural settlement dynamic and burial numbers in the associated Pantanello cemetery in the territory of ancient Metapontum, southern Italy (chart prepared by K. Sbonias, based on [10]).

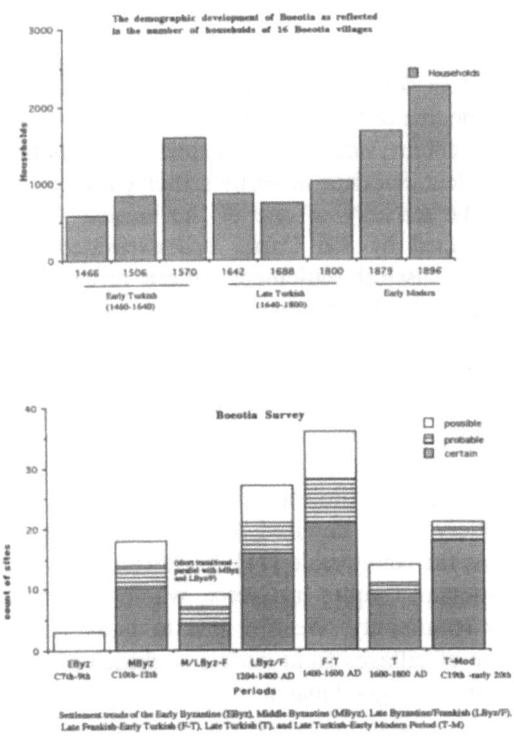

Fig. 9: A comparison of demographic developments in the province of Boeotia during post-Medieval times, Greece, based on Ottoman Imperial archives (upper chart, after Dr. M. Kiel), and archaeological surface survey (lower chart).

Probing the Secrets of the Subsoil using Geochemistry and Geophysics

Another way in which we can clarify the evidence gathered purely from surface artefact distributions, is to enlist the aid of several scientific techniques to investigate human activity traces in the surface and immediate subsurface deposits in a non-destructive and inexpensive way.

Human activity is associated with an immense accumulation of organic and inorganic residues - what today we would call 'pollution'. Ancient populations often used such waste materials to increase the fertility of their fields through the process of intensive manuring derived from household and farmyard rubbish. Adjacent to and east of the ancient Boeotian city of Hyettos, for example, lies an extensive plain. Mapping of surface artefacts across its surface shows a dense carpet of broken pottery that must reflect continous, largescale transport of city rubbish onto the plain (Fig. 10). Tony Wilkinson has shown that the radius of such manuring is a direct reflection of the population size of the settlement concerned (Fig. 11) [12].

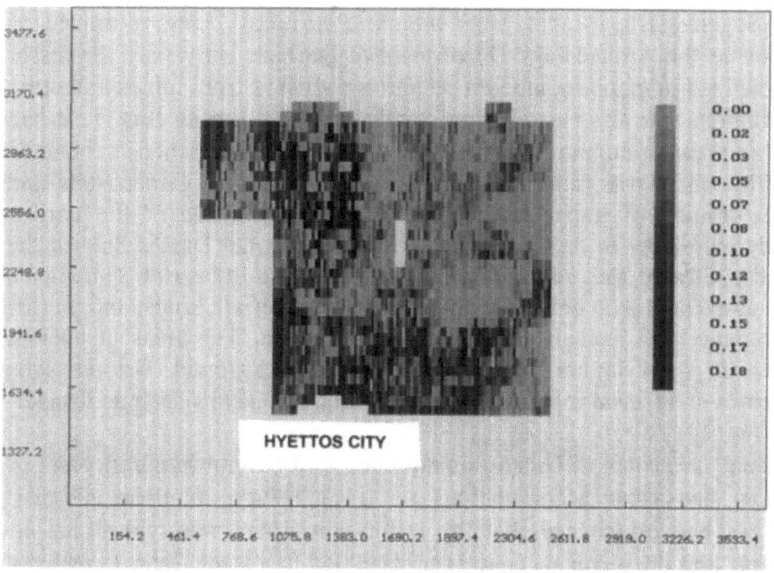

Fig. 10: The density per square metre of surface pottery in the immediate vicinity of the Greco-Roman city of Hyettos, Boeotia, Greece. The dense carpet of ceramics directly north-east of the town marks the main cultivable plain in its territory.(GIS by Dr. M.Gillings, author's data).

Clearly the inorganic broken ceramics have found their way into the soil alongside larger quantities of organic waste which have now been merged into the subsoil. This is the basis for a research programme into the soil chemistry of archaeological on-site and off-site sediments which we began in the 1980s on the

The approximate radius of significant field scatters surrounding archaeological sites in the Middle East (total sample: 19 settlements).

Settlement size	*Radius of scatter (km)*
Hamlets and farmsteads <1.5 ha	0.2–0.4
Villages 2–9 ha	0.6–1.0
Small town* 10–29 ha	1.3
Large town/city >40 ha	2.2–6.0

*One example only: site 48 in the North Jazira.

Fig. 11: Correlation between settlement size and manuring radii in the ancient and prehistoric Middle East (from [12] Table 1).

Boeotia Project, in collaboration with Professor Brian Davies of Bradford University Environmental Sciences Department, work which is currently being continued by Neil Rimmington in the Archaeology Department at Durham University. Professor Davies had specialized in mapping modern pollution pathways from mines and other industrial installations into the surrounding countryside, and during this work had discovered that all areas of human habitation showed soils with enrichened chemical elements - whether there was industrial waste or not. Analysing garden soils and comparing their content of trace element metals against the age of the garden revealed a steady increment in these chemicals over time - what Davies dubbed the 'habitation effect'. Clearly this research suggested that pre-Industrial settlements ought also to be associated with enhanced soil chemicals of certain kinds, and on this basis our programme with ancient Greek settlements began. The areas of ancient offsite manuring do show higher levels than regional background, but the soils across ancient cities show even more dramatic rises in trace metals such as Copper, Lead and Zinc (Fig. 12) [13].

Even small farmsites of Greco-Roman date, with several hundred years of occupation, show detectable accumulations of trace metals in their subsoils. However, once we had established that the Habitation Effect was typical for all periods of the past, and the elements concerned were tied in a stable form to the clay fraction of the soil, the mere confirmation of Davies' hypothesis was replaced by a deeper search for patterning in the dispersal of chemically-defined waste on ancient sites. A current focus of our research will illustrate the potential of such work. Site VM70 in the Boeotian Valley of the Muses is a typical Classical Greek farmsite. Its surface ceramic plot (Fig. 13) reveals an inner dense zone and a surrounding high-density zone (we call this a 'halo'), whilst the inner zone also had a dense area of rooftile indicating that here lay the main farmbuilding.

Intensive geophysical investigations using a resistivity technique (by Chris Gaffney, Mark Gillings and Neil Rimmington) showed an excellently clear plan (Fig. 14) focussing on a large, square, multi-roomed farmhouse in the very centre of

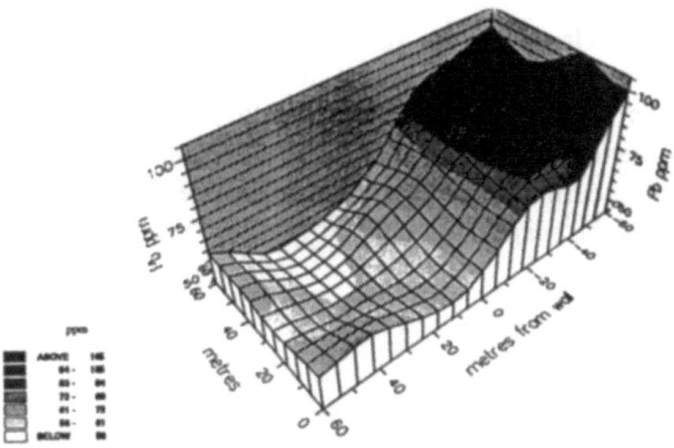

Surface soil lead concentrations at Thespiae.

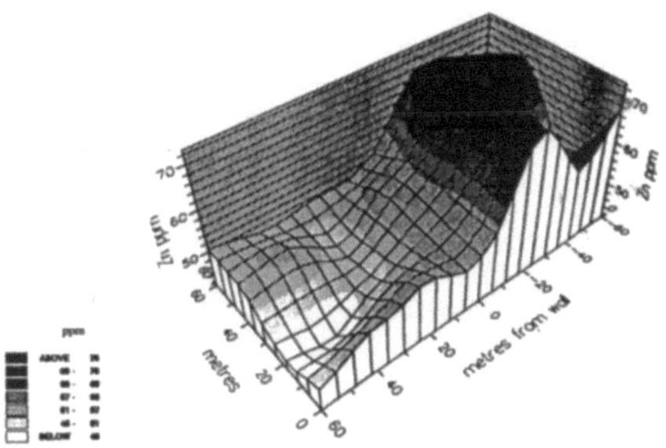

Surface soil zinc concentrations at Thespiae.

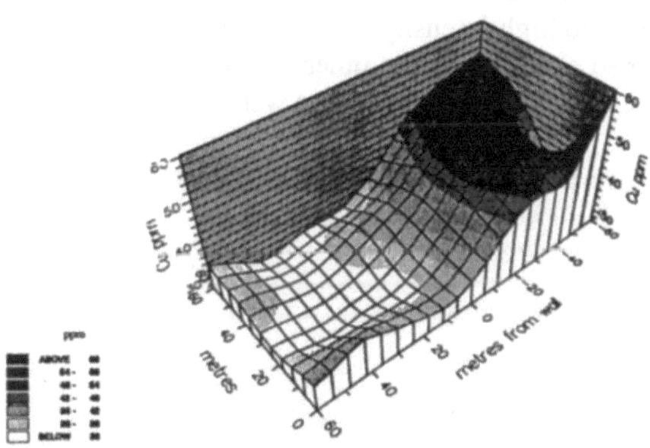

Surface soil copper concentrations at Thespiae.

Fig. 12: The transformation in the values of trace metals in surface soil along a transect running from outside the ancient city wall to inside the city, at the Greco-Roman town of Thespiae, Boeotia, Greece (prepared by N. Rimmington from data of Prof. B. Davies, cf. [13]).

the tile and densest ceramic spreads. However the resistivity plot also revealed evidence for a large enclosure ditch on all sides of the farm, defining a farmyard or garden zone, which had already been suggested through the 'halo' of high density ceramic on the surface. Trace element analysis of the soils using ICP by Neil Rimmington (Fig. 15) shows a high accumulation of lead over the farmhouse and high phosphate readings in and around the suspected enclosure ditches.

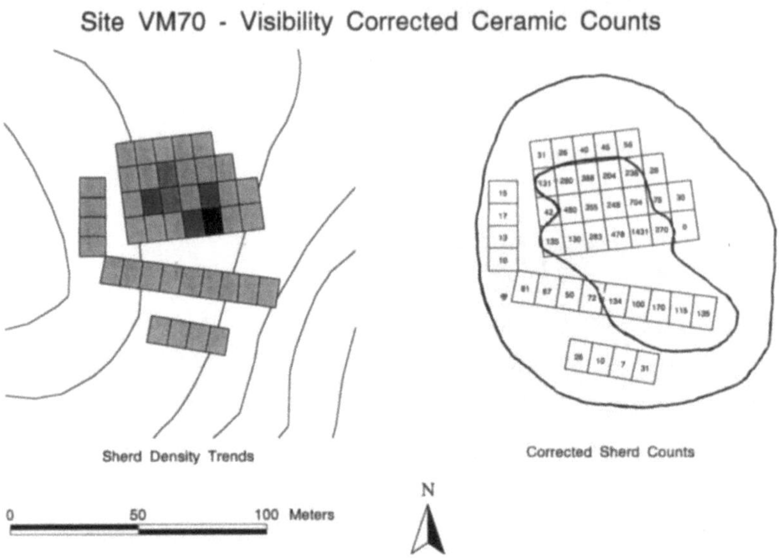

Fig. 13: The surface site VM70 in the Valley of the Muses, Boeotia, Greece. The surface ceramic density across the site, corrected for visibility, reveals a site core of very high values and a zone of medium to high intensity around it. A separate plot of tile debris on the surface also correlated closely with the inner sector, suggesting that the main dwelling structures of this Classical Greek farmhouse were located here.

Soil Resistivity Survey at VM70

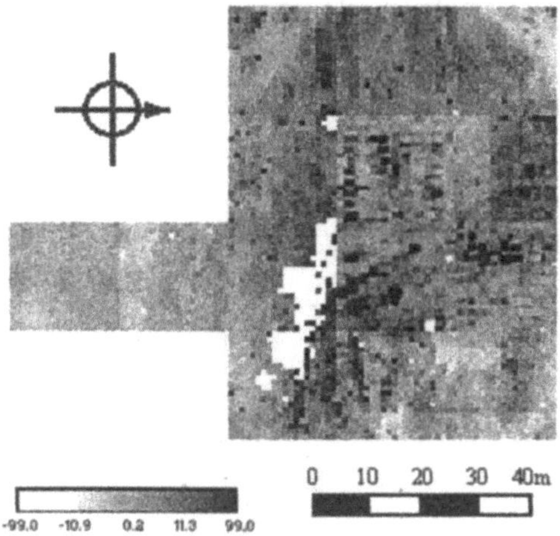

Interpretation of Results

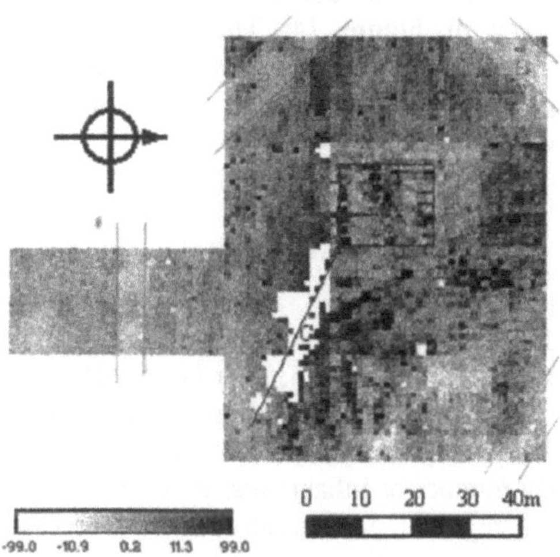

Fig. 14: A geophysical (electrical resistivity) survey carried out across site VM70. The elaborate rectangular and multi-roomed farmhouse in the core of the site is readily distinguishable, confirming that the inner highest ceramic density zone with its tile focus indeed overlies the dwellinghouse. However there are also strong indications on the south, south-west, north-east and north-west corners of the plot of an enclosing ditch system; this would very closely match the outer boundaries of the next zone of medium to high surface ceramics and would correspond to a farmyard or garden zone (research of C.Gaffney, M. Gillings, N. Rimmington for the Boeotia Project).

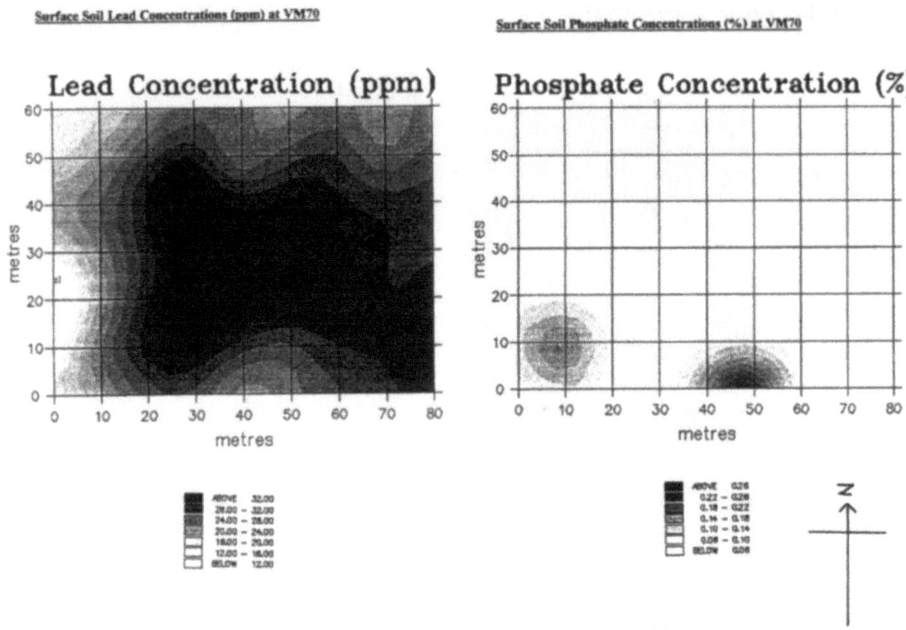

Fig. 15: Soil values for trace metal lead and phosphate across the main rectangle covered by the electrical resistance survey of Figure 14. The focus of lead corresponds to the location of the farmhouse, whilst the phosphate appears to be associated with the general zone of the putative enclosure ditches surrounding the settlement (research by N. Rimmington for the Boeotia Project).

Historical Interpretation

At the conclusion of an intensive regional surface survey, one can create histograms showing the fluctuations of site numbers against time (Fig. 16; after [14]). Allowing for the variable sizes of sites and the problem of whether all settlements were contemporary within a given phase, and provided that care is taken in trying to convert the area of a surface site to a likely number of inhabitants, such histograms can be used to chart the ebb and flow of regional demography. Such exercises, especially if town is compared with country population, may only be providing general order figures rather than exact and indisputable numbers, yet they are probably the best source of information we have about the behaviour of regional communities on the long timescale.

Thus in Greece, survey has cut through a long dispute from the literary evidence concerning the prosperity or otherwise of the Early Roman period, by showing a massive depopulation of rural sites and shrinkage of city areas at this time (see Figures 17-18 for a rural area in the Argolid, after [15]; Figures 19-20 for the shrinkage of the Boeotian city of Hyettos, from my own project).

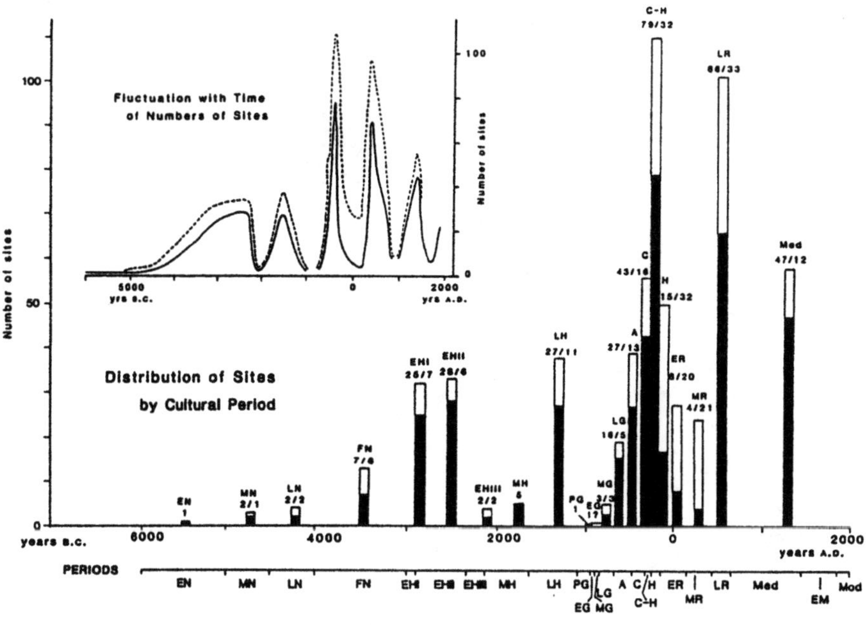

Fig. 16: Fluctuations in site numbers on the intensive survey of the S.W.Argolid, Greece (from [14] fig.4.4).

Fig. 17: Distribution of sites during the Late Classical-Early Hellenistic period on the S.W.Argolid Survey (from [14] fig.4.23).

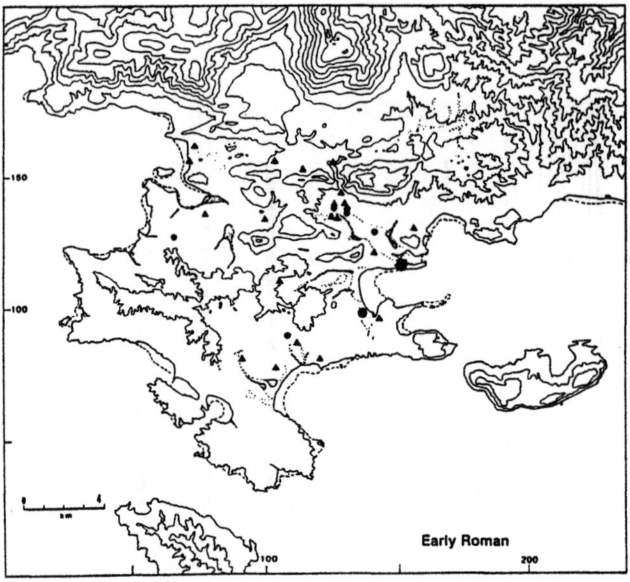

Fig. 18: Distribution of sites during the Early Roman period on the S.W.Argolid Survey (from [14] fig.4.25).

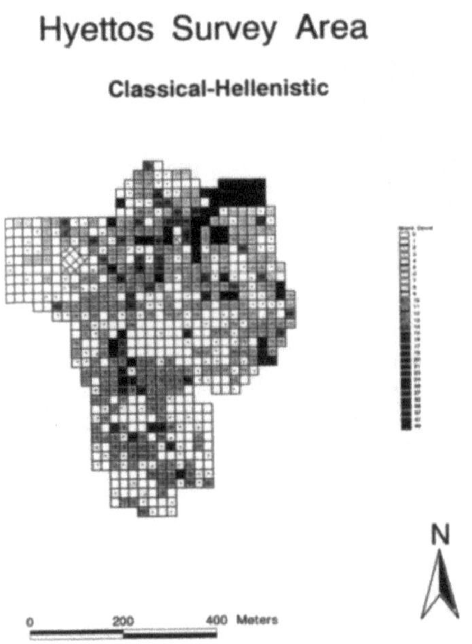

Fig. 19: Distribution of Classical-Early Hellenistic pottery across the surface of the city of Hyettos, Boeotia. The sample grid consists of over 600 units each 400 square metres in size (prepared by S.Fuller from the author's data).

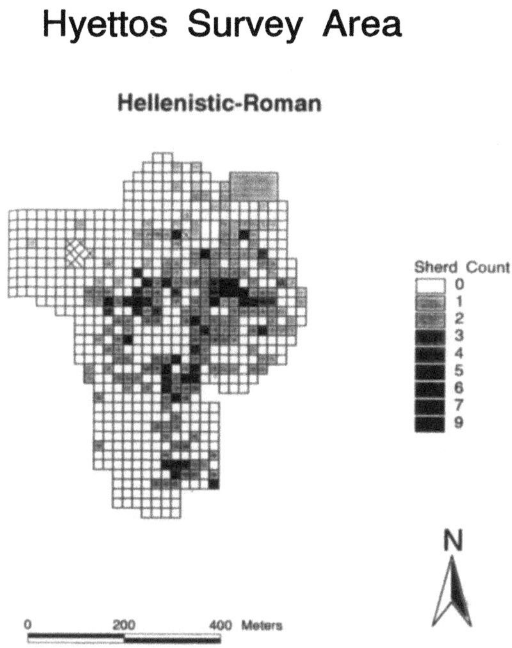

Fig. 20: Distribution of Late Hellenistic-Early Roman pottery across the surface of the city of Hyettos, Boeotia. A clear reduction both in the size of the occupied area and the intensity of activity seem indicated. (Prepared by S.Fuller from the author's data).

Further Interpretative Possibilities

In the late 1990's we are now in the situation in many regions of Southern Europe where it is possible, using regional surface survey, to compare and contrast the demographic curves of individual regions with each other, allowing an interregional perspective. Such a wider view is essential since we know that regions of pre-Industrial Europe did not exist in isolation, but were tied, in varying degrees, to political and economic, as well as demographic trends in other regions. Nonetheless, it seems clear that regions normally had strong internal development trends that interacted both positively and negatively with influences at the interregional level.

I recently undertook a preliminary comparison across the different provinces of Greece and the Aegean Islands of the timing of population and urban climax between 1000 BC and 700 AD [16]. Figure 21 shows the generalized results of this exercise. It proved possible to identify a limited number of models which could account for the differing growth trajectories of individual regions. Figure 22 shows the models which accounted for most of the variability in the data.

At the opposite end of the spatial scale, current research developments in surface survey are also focussing on variability, but this time of surface ceramic finds across each region. Thus Todd Whitelaw on the Kea Project [17] has examined the frequency of types of cooking, storage and tableware at Classical rural sites and compared them with a town site. On the Boeotia Project, Joanita Vroom [18] has been studying surface pottery from Medieval village sites and begun to relate the changing shapes to alterations in cooking and eating habits in High Medieval and early Post-Medieval Europe.

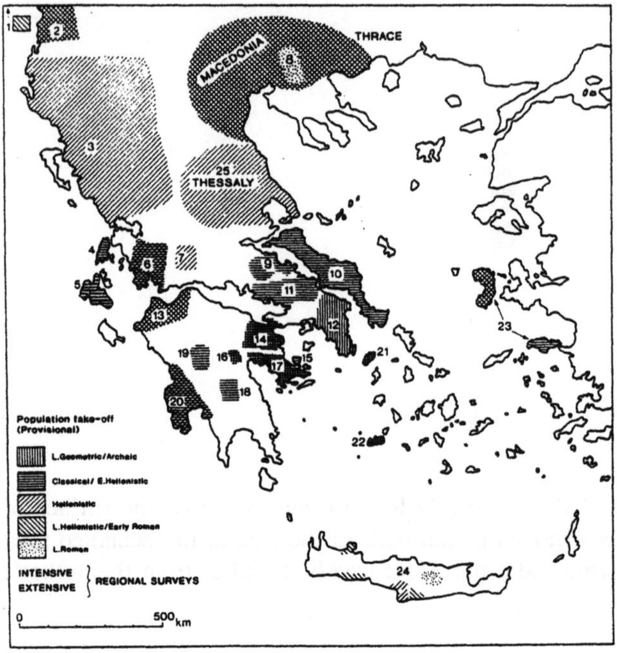

Fig. 21: The timing of demographic and urban takeoff in the different provinces of ancient Greece between Geometric and Late Roman times (from [16])

GIS and Virtual Reality Regions

With the rapid introduction of digitised maps and the use of Geographical Information Systems software to analyse and manipulate archaeological spatial data, the rich databases of settlement history that regional survey provide are being explored in dynamic and very novel ways. GIS allows a rapid calculation of problems that formerly involved long fieldwork, for example the relationship of a surface site to different land categories in its vicinity, the communication potential of a settlement location, the intervisibility of sites from each other. But via virtual reality software, we can take the digitised maps, add the discovered settlements, shrines and sanctuaries of any particular time in the past, plus our best reconstruction of land use and vegetation cover, and 'walk back into' a virtual reality survey landscape. Apart from the heuristic value of exploring the archaeological landscapes we have pieced together, as if we were ancient inhabitants, these new tools will prove of immense value in communicating the results of regional projects to the general public in the most accessible and enjoyable way.

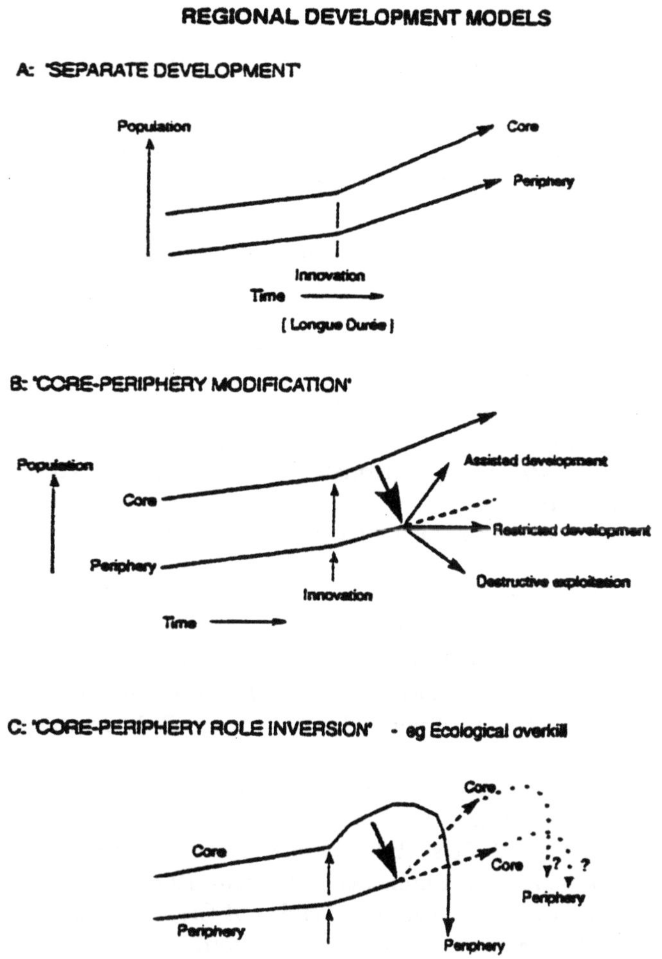

Fig. 22: Models for regional development in ancient and later prehistoric Greece. These models largely account for the variability in regional growth dynamics observable from current intensive and extensive survey data (from [16]).

Conclusion

Recalling my opening comments on the imaginative attempts of Classicizing painters to place us in past landscapes, current computer imaging has in some respects overtaken Poussin and Lorraine by allowing us to walk into and around Classical landscapes, whilst the rapid progress of field survey methodology itself has given us a far more realistic reconstruction of what those and other past landscapes actually consisted of.

References

1. H. G. Lolling, *Reisenotizen aus Griechenland 1876 und 1877* (Reimer Verlag: Berlin, 1989).
2. H. Lehmann, Geographische Zeitschrift 45, 212-228 (1939).
3. W. A. McDonald and G. R. Rapp (Eds.), *The Minnesota Messenia Expedition. Reconstructing a Bronze Age Regional Environment* (University of Minnesota: Minneapolis, 1972).
4. T. W. Potter, *The Changing Landscape of South Etruria* (Elek: London, 1979).
5. C. Renfrew and M. Wagstaff (Eds.), *An Island Polity. The Archaeology of Exploitation in Melos* (Cambridge University Press: Cambridge, 1982).
6. J. L. Bintliff and A. M. Snodgrass, Journal of Field Archaeology 12, 123-161 (1985).
7. J. L. Bintliff and A. M. Snodgrass, Antiquity 62, 57-71 (1988).
8. J. C. Chapman, J. L. Bintliff, V. L. Gaffney and B. Slapsak (Eds.), *Recent Developments in Yugoslav Archaeology* (British Archaeological Reports, International Series 431: Oxford, 1988).
9. B. Wells, C. Runnels, and E. Zangger, Opuscula Atheniensia 18(15), 207-238 (1990).
10. J. C. Carter (Ed.), *The Pantanello Necropolis 1982-1989. An Interim Report* (The University of Texas at Austin: Austin, 1990).
11. J. L. Bintliff, in J. L. Bintliff, & H. Hamerow (Eds.), *Europe Between Late Antiquity and the Middle Ages. Recent Archaeological and Historical Research in Western and Southern Europe*, 111-130. (Tempus Reparatum, BAR International Series 617: Oxford, 1995)..
12. T. J. Wilkinson, The Journal of Field Archaeology 16, 31-46 (1989).
13. J. L. Bintliff, B. Davies, C. Gaffney, A. Snodgrass, and A. Waters, in S. Bottema, G. Entjes-Nieborg, and W. V. Zeist (Eds.), *Man's Role in the Shaping of the Eastern Mediterranean Landscape*, (A.A.Balkema: Rotterdam, 1990).
14. M. H. Jameson, C. N. Runnels and T. H. van Andel (Eds.), *A Greek Countryside. The Southern Argolid from Prehistory to the Present Day* (Stanford University Press: Stanford, California, 1994).
15. M. H. Jameson, C. N. Runnels and T. H. van Andel (Eds.), *A Greek Countryside. The Southern Argolid from Prehistory to the Present Day* (Stanford University Press: Stanford, California, 1994).
16. J. L. Bintliff, Journal of Field Archaeology 24, forthcoming (1997).
17. T. Whitelaw, in R. Francovich and H. Patterson (Eds.), *Methodological Issues in Mediterranean Landscape Archaeology: Artefact Studies* (Oxford, in Press).
18. J. C. Vroom, Annual of the British School at Athens, forthcoming (1997).

Holography and Other Interferometrical Techniques

Holography in Museums

Vladimir B. Markov

Institute of Applied Optics
National Academy of Sciences - Centro Internacional de Fisica
10G Kudryavskaya, Kyiv, Ukraine - A.A. 4948, Bogota, Colombia
vmarkov@iao.freenet.kiev.ua - vmarkov@trauco.colomsat.net.co

Abstract. Some aspects of display and technical holography applications in museum practice are considered. Particular attention is paid to the quality of the reconstructed image, holographic non-destructive methods, memory, and image synthesis.

1 Introduction

Holography, as a powerful tool for 3D imaging, has attracted the attention of specialists since the early 60's, very soon after its re-invention by Yu. Denisyuk and E. Leith. The possibility to apply display holography as an optical twin of the object was emphasized in the first paper by Yu. Denisyuk[1]. However, it took a long time to develop the reliable technique and technology that enable one to record high quality holographic images[2,3].

The holograms of the individual museum items have been recorded in several laboratories around the world. However, only after multiple presentations at the exhibitions did it become evident that holographic techniques can be as helpful as the other optical methods systematically used in museums[4]. Later on, two trends were set apart for possible museum applications of holography[5]. Although the partition is rather conditional, the role of display and technical holography is nevertheless quite different. The holographic displays are mainly used for the presentation of museum items[6], while the methods of technical holography are exercised to study the conditions of preservation or restoration of the object[7,8], its identification[9], etc.

In this paper, several aspects of holography applications in museum practice are discussed. In particular, attention is paid to the reconstructed image quality, color imaging, methods of holographic non-destructive testing and holographic memory.

2 Holographic displays and their applications

A variety of motivations exists for holographic displays to be used in museum expositions. Some of those are: limited demonstration area; difficulties with real objects transportation; or precautions against harmful environmental conditions. As a result of these problems, very often the largest, invisible part of the collection is hidden in the museum depository. A partial solution to the problems of collection *latency* may be found by using the holographic copies of the real objects. Display holography can also be effective for high resolution imaging in archaeology[10].

Practical applications of display holography initiate certain demands from the experts. Their requirements are related mainly to the object's safety conditions at laser illumination and to the quality of the reconstructed image. In the latter case, such factors as holographic image observation conditions, its brightness, resolution, distortion, and color reproduction are of importance. Some of these parameters have been discussed before[11]. It was demonstrated that reflection hologram can simulate the conditions of observation, which are identical to the object itself. In particular, the imaging properties of the hologram, i.e. the *visibility zone* and *zone of vision* are close to real item perceiving. For example, an object measuring 40 cm x 40 cm x 40 cm, recorded onto the hologram measuring 60 cm x 80 cm can be observed on every side from 70° and will be entirely seen within an angle ~ 40° at a distance of 1.5 m from the hologram, as in a show-case.

Moreover, certain features of holography extend the capabilities of a museum item demonstration. For example, sequential recording of two different holographic images on one layer allows one to solve the problem of demonstrating items with two-sides (medals, coins, etc.)[5]. Another example is the reconstruction of the pseudoscopic image from the object as molds for casting. This particular technique was applied to retrieve the real images from some of the IV - IIIth c. BC terra-cotta casting molds[6].

Brightness of the holographic image is another important aspect, and it should be compared with the luminance $\mathbf{B}_o$ of the real object in exposition. The latter, according to [4], is acceptable if $\mathbf{B}_o \geq 70$ cd/m^2. The luminance of the holographic image $\mathbf{B}_h$ depends upon the conditions of reconstruction and characteristics of the hologram, i.e. illuminating light source intensity $\mathbf{E}_h$, diffraction efficiency η, etc. For a reflection hologram of $\eta \approx 25\%$ lit up with a halogen lamp ($\mathbf{E}_h \approx 700$ lux) the value of $\mathbf{B}_h \approx 60$ cd/m^2 is acceptable and is quite close to what is required. However a more essential criterion is the ratio between the luminance of the holographic image and the real object $\Im = \mathbf{B}_h/\mathbf{B}_o$. A standard level of museum object lighting is about 100 ± 50 lux; the value of $\Im$ can be made very close to that of the museum object.

2.1 Quality of the reconstructed holographic image

Characteristics of the light source used for holographic image reconstruction (**RLS**) strongly influence the quality of the latter. In particular, the holographic image quality depends upon the spatial position, angular dimensions and spectral bandwidth of the **RLS**[12,13]. The level of the image blur is proportional to **RLS** angular dimensions and depth z_o of the imaged scene. For a hologram with a depth of scene of about 25 cm illuminated by a halogen lamp with the luminous body $\delta x_r \approx$ 0.5 cm from a distance of 200 cm, the image blurring is relatively low ($\Delta x_i \approx 0.6$ mm) and does not reduce the image quality.

A stronger effect on the reconstructed image characteristics is associated with the spectral bandwidth of the light source. In this case, the image blurring depends upon the depth of the image, recording media thickness T, and grating spacing. For example, the blur of an image point with a depth of 25 cm can reach $\Delta x_i \approx 5$ mm (for $T = 7\mu m$, $\lambda = 570$ nm). This value of Δx_i considerably degrades the image quality and should be taken into consideration (Fig. 1). The **RLS** with a narrow spectral bandwidth reduces the spectral blur, however strongly increases the requirements to the uniformity of the emulsion, as any inherent non-homogenates become more visible.

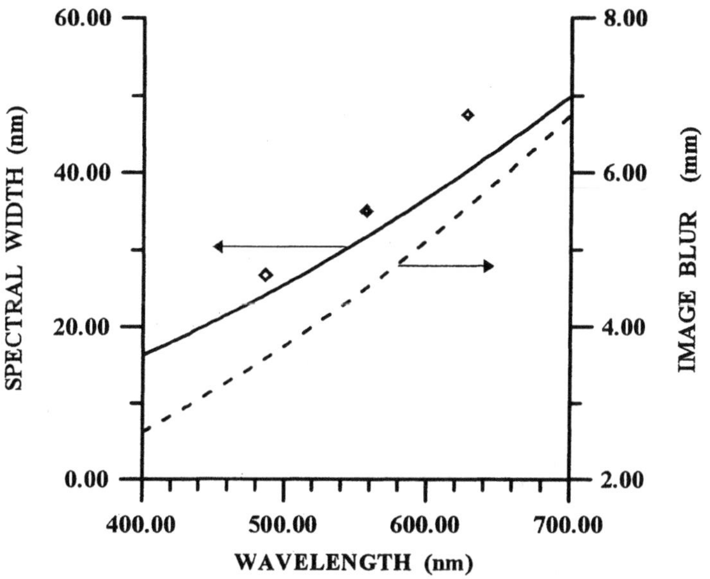

Fig. 1. Spectral width and image blur of an image from the reflection hologram with layer thickness 7 μk.

2.2 Color holographic imaging

Of special importance for museum applications is true color holographic image reconstruction, which has remained one of the critical problems for many years. As a matter of fact, not all the objects require color recording. Some of them, like gold items, are well reproduced by a monochrome hologram, if the spectral band of the diffracted beam is tuned properly towards the yellow part of the spectrum. Nevertheless, color holographic recording remains rather important.

Experimentally, color reproduction was a subject of investigation from the early beginnings of display holography[14]. Rather extensive study of this problem was made in [15], where different aspects, including recording wavelengths selection, chemical processing, and the noise in the recording media have been analyzed. Although technically the task seemed to be solvable, the basic problem was the lack of the panchromatic recording materials.

Recently produced PFG-03C silver-halide layer seems to be very efficient for the discussed purpose [16-17]. Then it was demonstrated[16] that sufficiently good quality color holograms can be recorded on this film using 488 nm, 532 nm and 488 nm laser lines with their intensity ratio as 1:0.85:0.65, respectively, at a total exposure ~10^{-2} J/cm^2. In combination with modified Holodev-602 developer and PBQ-Amidol bleach a color hologram with $\eta \approx 25\%$ was recorded.

The joint function of the angular-spectral selectivity $\eta(\theta, \lambda)$ was found to be a useful parameter to analyze the color hologram properties as it determines the off-Bragg replay characteristics. Fig. 2 illustrates, calculated in the frame of the coupled-wave equations[18], the spectral-angular response of the reflection grating. It shows that any deviation of the reconstructing (or observation) angle results in corresponding changes of the reconstructed wavelength, and vice versa. As a result, the balance of the colors in the reconstructed image will be violated.

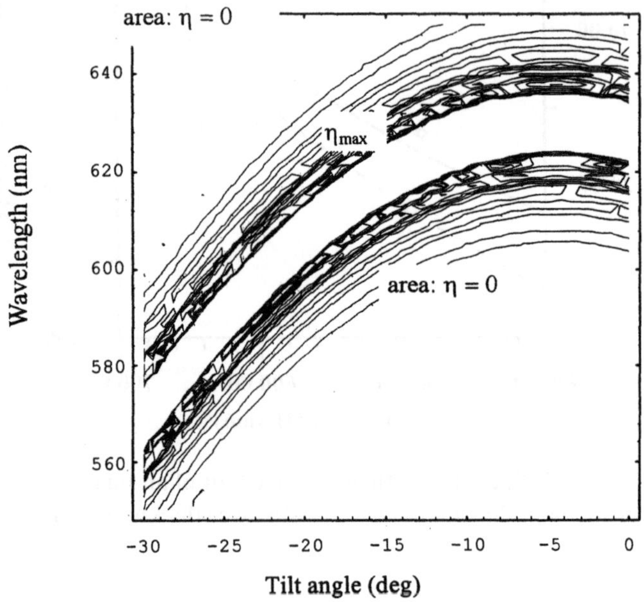

Fig. 2 Spectral-angular characteristic of the reflection hologram calculated for recording-reconstruction conditions at $\lambda = 633$ nm

3. Technical holography

3.1 Holographic non-destructive testing

Museum items are subjected to a continuous adjustment process related to variations in the environmental conditions. As a result, any existing or latent defects initiate further deterioration of the object. This very often happens with frescoes and icons in churches. Cyclic changes in humidity and temperature induce local defects in the plaster and its substrate, and can cause the object's degradation. The microscopic level of these processes is the main problem in monitoring them.

Holographic interferometry was applied for the detection of defects and stress analysis for different kinds of museum objects (see ref. in [11]). The most efficient one for this purpose seems to be electronic speckle pattern interferometry (**ESPI**). The advantage of this technique is that it can operate practically in real time, has image processing and gives high resolution for out-of-plane displacement. This is of interest, taking into account the number of items in the museum to be studied.

A detailed description of an experimental **ESPI** system for museum applications can be found in [8, 19, 20]. It consists of optical and electronic parts, as well as specially elaborated software. The method is based on the successive acquisition of "holographic" images with a standard CCD camera. Image sequences that correspond to the non-disturbed and disturbed states of the object are taken by this camera supported by an image processing board, and then stored in the computer memory. The algorithm processing of the stored images is based on a standard phase-stepping technique[21]. The sensitivity of the developed system to out-of-plane displacement measurements is about 0.15 μk.

An example of the **ESPI** system operation for detection of defects in an icon from the XVIII century is shown in Fig. 3. The icon is a typical multi-layer structure. It consists of a wooden support, which was coated with several priming sub-layers and has been gilded over the entire panel area. The gold surface is partially painted with oil-paint. The problem of such a multi-layer structure is connected with the formation of the local detached regions between the priming layers and wooden panel on the one hand, and between the gilded surface and the sub-layers on the other.

Relatively low temperature variations of the icon's surface, initiated by an IR-lamp during 30-40 sec, were found to be sufficient to obtain observable results. The measured variations of the temperature on the surface of the object did not exceed 0.5°C. Fig. 3,a shows a fragment of the icon imaged through the CCD camera. The IR-lamp thermal action on the object initiates the local changes of the interference pattern (Fig. 3,b). The observed non-uniformity in the fringes can be classified as

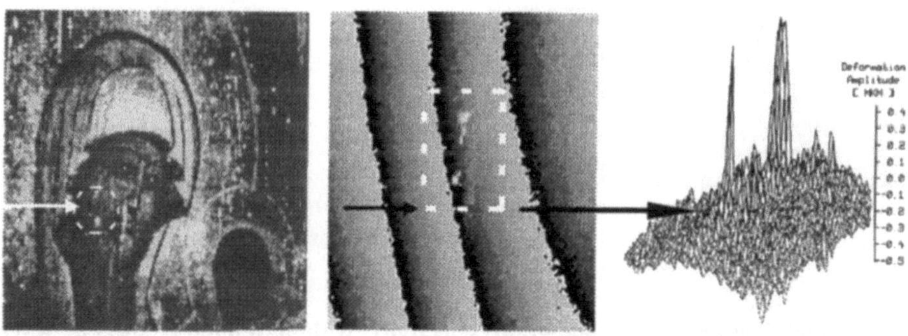

a) b) c)

Fig. 3(a,b,c). A fragment of a XVIIIth c. icon imaged through CCD camera (a), its interferogram with defect localization (b) and the surface detachment area (c).

out-of-plane displacement, which in the case under consideration can be attributed to local detachments (up to 0.7 μk) between priming and gold layers (Fig. 3,c).

3.2. Holographic memory system.

The development of a high density storage system for the description of museum items (a so-called "museum passport") is an important task. Computer systems are widely used for this purpose. However, taking into account the bulk of information to be recorded, it is worth considering more efficient systems, one of which could be the holographic memory. According to up-to-date ideas, the highest level of information storage density is accessible by using volume holography[21]. Different approaches have been suggested for this purpose, mainly based on dispersion (angular and spectral) features of a 3-D hologram. For example, the possibility of storing 5000 holograms in a 1 cm thick $LiNbO_3$ crystal with a total capacity of about $1.5*10^9$ bites (for image recording) was recently demonstrated[22]. One of the main problems here is the cross-talk during the information recording and retrieving.

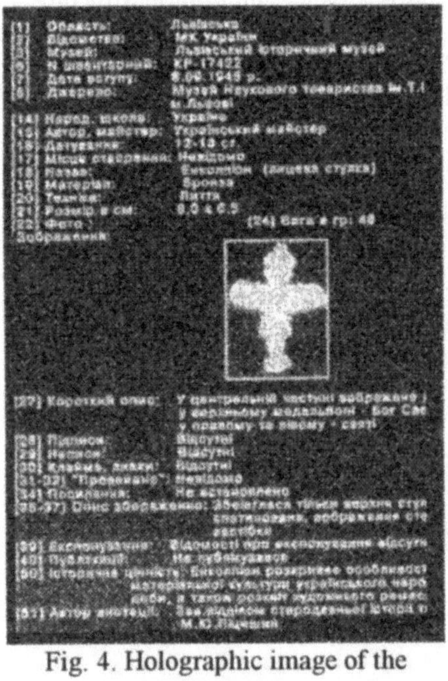

Fig. 4. Holographic image of the museum passport

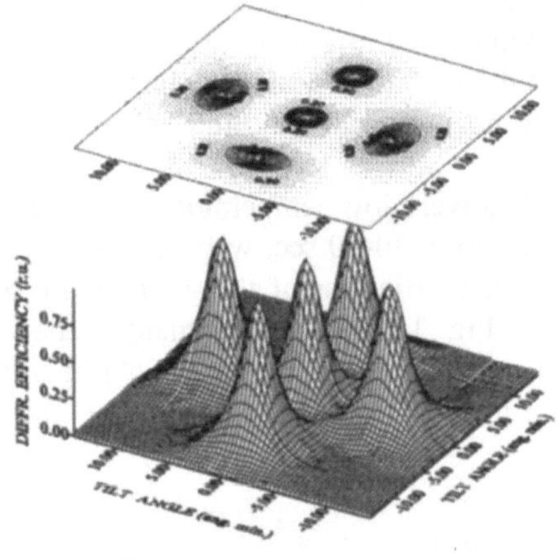

Fig. 5. Information multiplexing through angular speckle-selectivity

The modification of 3D hologram selectivity with a reference speckle wave (**RSW**) was analyzed in [23]. The finite size of the individual speckle leads to additional selectivity related to the spatial mismatch (lateral and longitudinal) of the recorded and reconstructing speckle fields. Both of these types of selectivity can be described by the normalized auto-correlation function of **RSW**, determined by an *average* speckle size in both lateral Δ_T and longitudinal Δ_L directions, respectively. As a result, conditions can be fulfilled when several images are superimposed by changing the relative position of the hologram and speckle structure. Fig. 4 illustrates a typical image of a museum passport reconstructed from the volume hologram with speckle-shift selectivity.

The angular selectivity of the 3D **RSW** hologram also differs from that recorded by plane waves[24]. In particular, its value is determined not only by the cross-grating period, but also by the de-correlation of both amplitude and phase of the stored and reconstructing speckle fields. As a result, the conditions can be realized when the so called "speckle angular selectivity" is stronger than the cross-grating angular selectivity. Moreover, contrary to the traditional angular selectivity, which exists only in the direction of dispersion, in the case of **RSW** it takes place for any direction of the hologram tilt/rotation. The illustration of the above said is presented in Fig. 5, where the sequence of the individual holograms is reconstructed with a recording medium tilt around two orthogonal axes.

3.3 Holographic image synthesis

Holographic 3-D image synthesis could also be of special interest. This technique was suggested more than 20 years ago[25-27]. The possibility of large size holographic 3-D image simulation was demonstrated in [28]. Much later, this technique was improved by using computer synthesized images with their transfer into the hologram through a spatial light modulator. Finally, a real time electro-holography 3-D imaging system was suggested and demonstrated[29]. Although the quality of the simulated image still needs to be improved, the advantages of this method are quite promising.

In considering possible applications of holographic 3-D image synthesis, we should not forget several options where this technique could be efficient and important in museum practice. First of all, it is a well known fact that there are objects in the museum collections which, for different reasons, can not be exhibited, many of them being either completely lost (destroyed) or strongly damaged. In the case where the real object does not exist anymore, it could be very useful to compose its optical copy. At the same time, the application of the holographic technique 3-D image synthesis makes it possible to join the existing part of the object with the holographically originated non-existing part. This technique can help to present a real object together with a holographically simulated image. These two parts will complement each other, generating an illusion of the whole item.

As a matter of fact, the above described technique of holographic image generation and its following integration with the still existing part of the object can be very helpful for the experts and specialists who are working with the restoration of museum items. For them, a generated 3-D holographic image and its following combination with an existing part of the object can serve as important additional information for the most correct and optimal method of restoration.

Other applications of holographic 3-D image synthesis could be in the area of large size space observation and for simulating the images of now non-existent ancient monuments, architecture memorials, etc. It is evident that during the centuries of civilization, many of these monument have been badly destroyed or damaged as a result of natural disasters or catastrophic events. Also, the latest tendencies in the architecture of urban style have brought a lot of transformation to current cities, compared to what they were like some centuries before. All this results in very strong changes in the city "environment", making it practically non-reproducible with existing techniques.

Similar problems exist in archaeology. The excavation of original monuments, basically done with the purpose of more detailed study, often comes along with certain changes in the monument itself. Sometimes these changes reach the level where it becomes impossible to reconstruct the object in its initial form, not only as a result of technical, but also for practical reasons. Of course, archeological studies in specially preserved places that belong to the national or world recognized list of heritage allow the conservation of the monument. However, we have to recognize that the problems related to the reconstruction of the monuments or fixing their present state are basically connected with the cost of the work to be done. A more complicated situation is observed when a monument was discovered inside of an urban area and, for practical reasons, cannot be reconstructed without damaging the existing surrounding structures. A positive and impressive example in solving such a problem is the Jorvik Viking Center, which has been created on the site of excavations in downtown York (UK) and does not negatively affect the city.

Thus, the problems discussed above demonstrate that a field exists where the replacement of a large size object through its 3-D model seems to be very convenient. Several possibilities can be suggested for this purpose, such as: big size posters; scaled models; and virtual reality systems. However, in one case it is just a 2D image or a relatively small model not wrapped up directly into the space of the natural "environment" of the real object. In the other case (like virtual reality) it is too individualized. So, an efficient solution to this problem could be a large size holographic panel with a synthesized 3-D image. The aspects related to the simulation and reconstruction of the large space scene with holography have been discussed in several papers. For instance, in [30,31] it was suggested to use a hologram for the presentation of space that existed about 2 thousands year ago and is not related to any of the previously existing objects.

In our case, the general idea is as follows: Let us consider an archaeological site such as Khersonessus Tauricus, the Greek city that is known to exist from IV century BC. After more then 150 years of study, archaeologists have been able to "open" the basic level of the city as it was in III-IV century AD. The complete reconstruction of the resort-like Khersonessus, or even some of its basilicas, would be extremely costly, and is not feasible in an observable period of time. However, using the above described principles of holographic 3-D imaging, it is possible to simulate the view of the whole city, its parts or a specific monument, not interfering much with real space. This is demonstrated in Fig. 6, where the holographic "screen" with a simulated image of certain area is shown. In the case under consideration it is used to reproduce the view of one of the destroyed basilicas. Taking into account that the contemporary holographic techniques and technology allow one to create the displays with high brightness and low noise, the discussed method can be used to simulate the full scale image of the initially existing monument, overlapping it with an existing part of the monument in its real environment. This will strongly increase the visual effects when perceiving reality.

4 Conclusions

Some aspects of holographic applications in museum practice are discussed. In particular, attention is paid to what are called display and technical holographic methods. In the case of display holography, the image quality can be substantially affected by the characteristics of the reconstructing light source. The true color holographic image, being very important for the demonstration of museum items, should be properly reconstructed. Correct conditions for the illumination of these types of hologram can be determined through the joint function of the angular-spectral selectivity.

In the case of technical holography, holographic interferometry, memory, and 3D holographic image synthesis are of special interest. The detection of an invisible defect on an XVIII icon is given as an example. High sensitivity of the described system to measure out-of-plane displacement allows the detection of microscopic cracks and other defects. It made it possible to detect the de-bonding between the priming layers and the support on the level of about 7 μk. Taking into account the

Fig. 6. A photograph of the excavations of the ancient Khersonessus (Crimea, Sebastopol) with a possible holographic simulation of a basilica of that period.

non-contact character of the tests and complete automatic control of the measurements, the application of this technique in the museum practice seems to be very useful and promising.

Recent developments in the holographic memory system and image synthesis are also presented and their possible applications are discussed.

References

1. Yu. Denisyuk, Docl. Acad. Nauk., **144**, 1275 (1962).
2. Yu. Denisyuk, Zh. Techn. Phys., **48**, 1683 (1978).
3. N. Phillips, Proc. SPIE, **532**, 29 (1985).
4. T. Brill, *Light. Its Interaction with art and antiquities.* Plenum Press, N.Y. (1980)
5. V. Markov, I. Yavtushenko, G. Mironyuk, *Holography and its Applications in Museum Work*, UNESCO Publ., 1984.
6. V. Markov, Neue Mus Kunde, **2**, 107 (1989).
7. S. Amadesi, F. Gori, R. Gralla, G. Guattari, Appl Opt., **13,** 2009 (1974).
8. P. Boone, V. Markov, Studies in Conservations, **40**, 103 (1995).
9. J. Duvernoy, D. Charraut, Opt Acta, **24**, 795 (1977).
10. D. Vukicevic, G. von Bally, W. Sommerfeld, in *Optics for protection of man and environments against natural and technological disasters* (Ed. G.von Bally, H. Bjelkhagen). Elsevier Sc. Publ. (1993).
11. V. Markov, Opt.&Laser Techn. **28**, 319 (1996).
12. G. Ross, J. Watson, Proc. SPIE, **1732**, 198 (1992).
13. R. Meier, J. Opt Soc Am., **55**, 987 (1965).
14. R. Collier, C. Burkhart, L. Lin, *Optical Holography*. Acad. Press, N.-Y. (1971)
15. P. Hubel, *Color reflection holography.* Ph.D. Thesis, The Univ.of Oxford (1990)
16. G. von Bally, F. Dreesen, E. B. de Haller, V. Markov, A. Roshop, Tech. Phys Lett. **21**, 667 (1995).
17. H. Bjelkhagen, D. Vukicevic, Proc. SPIE, **2333**, 34 (1985).
18. V. B. Markov, A.I. Khizhnyak, Techn Phys Lett. **22**, 18 (1996).
19. P. Boone, V. Markov, N. Burykin, V. Ovsyannikov, Proc. SPIE, **2648**, 596 (1995)
20. G. Gülker, K. Hinsch, C. Hölscher, P. Meinlschmidt, K. Wolf, in *Optics for protection of man and environments against natural and technological disasters* (Ed. G.von Bally, H. Bjelkhagen). Elsevier Sc. Publ. (1993).
21. R. Jones, C. Wykes, *Holographic and speckle interferometry* (Cambridge U.P., 1983).
22. D. Mok Opt. Lett. **18**, 915 (1993).
23. A. Darskii. V. Markov, Opt. Spectr. **65**, 661 (1988).
24. A. Darskii. V. Markov, Proc. SPIE, **1238**, 30 (1989).
25. R. V. Pole, Appli. Phys. Lett. **10**, 20 (1967).

26. J. T. McCrickerd, N. George, Appl. Phys. Lett. **12**, 10 (1968).
27. D. J. De Bitetto, Appl. Opt. **8**, 1740 (1969).
28. G.Newswanger, C. Outwater, Proc. SPIE, **523**, 127 (1985).
29. S. Benton, Proc. SPIE, **IS-8**, 247 (1990).
30. P. Dawson, Proc. SPIE, **1600**, 149 (1991).
31. M. L. Jepsen, P. Dawson, Proc. SPIE, **1732,** 387 (1992).

Holography: A New Technology in Cuneiform Research

W. Sommerfeld[1], G. von Bally[2], F. Dreesen[2], and A. Roshop[2]

[1] Dept. Ancient Near Eastern Studies, University of Marburg, Wilhelm-Röpke-Weg 6F, D-35039 Marburg, Germany
[2] Laboratory of Biophysics, University of Münster, Robert-Koch-Str. 45, D-48129 Münster, Germany

All systems of writing which are now in use worldwide are two-dimensional in nature. The information is provided by a difference in colour, most commonly black on white.

Few people are aware that the triumph of 2 D-scripts came about rather late in the history of writing. The invention of writing was a major step in the process of civilization, but in the beginning only 3 D-scripts were used. The writing materials like clay, stone, or metal are monochrome, and the information about language expressed in writing is not indicated by any difference in colour, but by the structure of the surface.

The most prominent example of this kind of inscription is cuneiform script. This system of writing — the very first one in the history of mankind — was established more than 5000 years ago, and it was in use for more than 3000 years (the latest cuneiform tablet was written in about 100 A. D.)[1].

Fig. 1 shows a photograph of an Old Assyrian cuneiform tablet; its age is about 4000 years. It is a protocol of a court decision, settling the quarrelings between two business enterprises[2]. The wording of the decision was written down on a tablet which was surrounded by this case. The case bears the names of the persons involved and impressions of their personal seals. The case was destroyed to obtain access to the tablet. Afterwards, the fragments were recombined. The difference in colour (in grayscale here in the black and white photograph) of the lower right part was the result of a different temperature while being fired in a kiln.

The writing material (i. e. clay) was very cheap and easy to inscribe. Consequently, cuneiform was used to issue documents which deal with nearly all aspects of everyday life as well as with intellectual matters. Thus, the cultural heritage of cuneiform tablets covers an enormous amount of economic and business documents, letters, royal inscriptions, and literature of every kind. These texts provide us with most detailed information about the cultures of the Ancient Near East.

Until now, up to one million cuneiform tablets have been excavated. They are now housed in hundreds of museums and collections worldwide. Moreover, it is to be expected that in the soil of the Near East there are still up to a 100 million tablets waiting to be discovered.

Fig. 1. Photograph of a fragmented case of an Old Assyrian clay tablet with cuneiform inscriptions and seals

This paper will present:

1. the problems that modern scientists face when dealing with cuneiform tablets;
2. the common methods of documenting these inscriptions and their shortcomings; and
3. based on the results of the research program on "archaeoholography" at the Laboratory of Biophysics, University of Münster, Germany, the advantages of the methods of holography and the fundamentally new possibilities for cuneiform science.

How were cuneiform tablets inscribed?

The elements of the script are rendered by a specific structure of the surface. The ancient scribe held a tablet, a pillow-shaped piece of clay, in one hand and a stylus in the other. Any triangular instrument with sharp edges could be used as a stylus. To produce a wedge, the scribe pressed the stylus into the clay, the top of the stylus slightly deeper than the bottom. This way the typical cuneiform shape of a wedge appears, i.e. a 3 D information carrier is formed.

According to the anatomy of the human hand, such a wedge could be impressed in four different directions: horizontal, vertical, and diagonal. Every cuneiform sign is composed of these basic elements.

The simplest signs are identical with just such a single wedge, but there are also very complicated signs, composed of up to 30 wedges (Fig. 2). Altogether, the cuneiform writing contained several hundred signs, each of them indicating different syllables or complete words.

There are many examples of signs which look very similar and therefore can be easily confused (see Fig. 3).

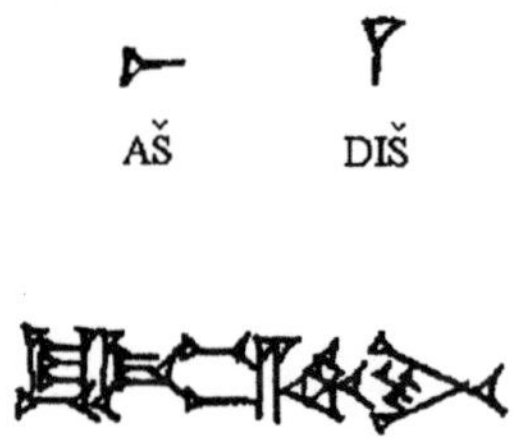

Fig. 2. Simple (up) and complicated (low) cuneiform signs

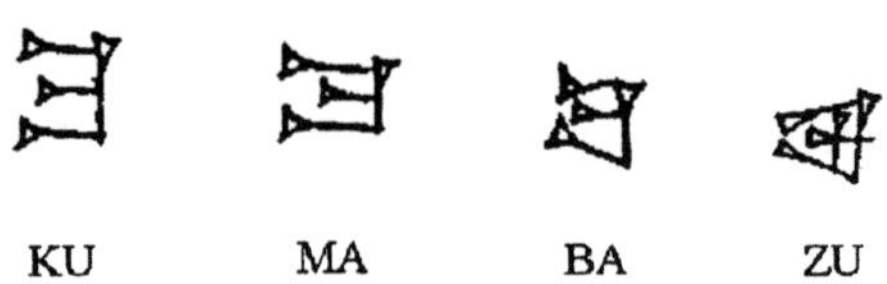

Fig. 3. Cuneiform signs of similar shape but different meaning

How are cuneiform texts read?

To identify a certain sign, one has to recognize the elements, i. e. the wedges. Since there is no difference in colour, identification depends mainly on the way the light falls on the tablet thus causing contrast by shadowing.

If the light source is centered above the tablet and if the light is directed on the centre of the tablet, there is no shadow at all. A nice tablet can be seen, but it is almost impossible to identify the different elements of the script. If the light impinges from the left, one may perceive a distinct shadow which allows the easy identification of the vertical wedges, but there is no contrast for the horizontal ones. Therefore, one fails to recognize their number and arrangement. If the light is directed from the top of the tablet, the situation is vice versa: one can see a good shadow for horizontal wedges, but it is difficult or even impossible to recognize the vertical elements. Therefore — in order to take photographs — the source of the illumination is usually fixed at an angle of 45 °, which causes a reasonable optical contrast for horizontal and vertical wedges.

When wedges are impressed only superficially or when the surface of a tablet is not well preserved and the edges of a wedge are worn, it is sometimes extremely difficult to identify wedges and signs.

Thus, the editor of a cuneiform tablet may easily confuse signs which leads to the misinterpretation of words or even sentences. If key words are misread, the meaning of an inscription may be entirely mistaken. A good deal of work as a cuneiform scientist is to correct the misreadings of colleagues. But if there is an idea of how a certain reading should be improved, it has to be checked on the original tablet to confirm this supposition. Consequently, since cuneiform tablets are spread in museums all over the world, this generally requires an enormous investment of time and money.

This background information is necessary to understand what standards have to be achieved to obtain an appropriate documentation of cuneiform tablets or 3 D-inscriptions in general, and to assess the advantages of holography.

For a documentation of cuneiform inscriptions that pretends to render all important information, two different kinds of methods have been in use up to now:

1. technical means of reproduction (like photography, digital CCD-recordings, 3 D-scanners, or casts),
2. manual methods (i. e. hand-copies of inscriptions).

1. (a) The technique of documentation best established is photography, but there are several major disadvantages:
   i. The main problem is that the illumination is fixed in a certain direction, usually at an angle of 45 °. The result is a specific contrast which cannot be changed. Some elements are clearly to be seen, while others are very faint, or even escape identification. Thus, only a certain amount of the information of the original tablet is recorded by photographs.
   ii. If the structure of an inscription is damaged, the problems increase considerably. It is often difficult to decide whether scratches visible on the surface of a tablet are part of a wedge or sign or whether they are only physical irregularities of the tablet like small holes etc. without significance. A scientist can only reach a decision when he obtains a very good contrast of the details. This is because wedges and scratches are not identical in their microstructure.

       Since photography can render a document only in a specific position with fixed illumination, it is impossible to change or improve the contrast. Thus, it is very difficult to distinguish between wedges and scratches on photographs. In some cases, photographs even turn out to be useless.

       In short, it is often possible to identify damaged wedges and signs on the original tablet, but not on the photograph of that tablet.
   iii. Finally, the objects bearing these inscriptions are also 3 D themselves. Cuneiform tablets usually are shaped like a pillow, and since the depth of focus is limited, photographs always have distortions, sometimes to an extent that excludes legibility.

Fig. 5. High-resolution holographic reconstructed images of a clay tablet with cuneiform inscriptions. Both images are recorded from the same hologram showing different perspectives as depicted by the different positions of the shadowed parts, thus demonstrating the three-dimensional recording capability.

Since the objects bearing 3 D inscriptions are 3 D themselves, distortions can be considerably reduced when tablets are recorded by holograms. Holography has limits concerning the depth of focus too, but this technique is more appropriate to render curvatures than any other one. It is even possible to look at the different edges of a tablet while the tablet as a whole is still visible.

2. The second advantage of holography is high resolution – more precisely: high data storage capacity.
   It is very important to inspect the microstructure to identify traces of wedges and signs. Furthermore, there are cuneiform tablets with extremely small inscriptions. The height of the cuneiform signs may be as small as one millimeter, thus the ancient scribe must have had very good eyes. It is impossible to read such a tablet without a strong magnifying glass. This can easily be used on a holographic reconstructed image (see Fig. 6).
   The lateral resolution of holographic reconstructed images which are available now goes down to about 3 μm (Fig. 7). Thus, such a hologram can be looked at with a magnifying glass or even a microscope to get a more distinct picture of the details.
   It can be estimated that thousands of micrographic photos of one tablet are required to get the same amount of information that is recorded by a single hologram.
   Furthermore, it is possible to make photographic documentation as well as digital CCD-recordings from holograms. Thus, they can be used as a source for image processing with the computer and for digitalized data bases [5].
   To judge the quality and reliability of hologram recordings, an original tablet, a very good photograph, and a hologram were compared by several cuneiform experts. For the test, a tablet was selected where the surface was badly damaged and therefore the cuneiform inscription was very difficult to read. It turned out that on the photo, about 1/3 of the traces of wedges

Fig. 6. Inspection of a holographic reconstructed image of the cuneiform tablet of Fig. 1 through a magnifying glass.

and signs could not be identified; it was impossible to distinguish between wedges and scratches. On the other hand, all these difficult details could be recognized clearly on the hologram. A careful comparison of every element on both the original tablet and the hologram proved that the original tablet does not reveal any information of the cuneiform inscription that is not clearly recorded by the hologram as well.
In general we may conclude: the more serious the difficulties of cuneiform tablets, the greater the advantages of holography.

3. Holograms can be copied [6].
Due to a specific recording technique the copies themselves may show an even higher contrast in the reconstructed image than the "master hologram" taken from the original. Thus, the creation of an archive seems to be possible.

4. Furthermore, an unparalleled advantage of holography is the possibility to (holographically) recombine several fragments of one and the same tablet in 3 D.
The original fragments do not have to be brought together for this purpose. Fitting tests and holographic recombination can be done using holograms of the original fragments (Fig. 8).

Fig. 7. Holographic reconstructed image of a resolution test chart demonstrating microscopic resolution (group 7.3 $\hat{=}$3.1 μm lateral resolution).

Fig. 8. Holographic recombination of a fragmented cuneiform tablet. The three-dimensional high-resolution holographic reconstructed image of the complete clay tablet (low) is created by using two independently recorded holograms of the fragments (up). The original fragments themselves have not been brought together; they may even no longer exist when the holographic recombination is made from the holograms of the fragments.

There are more promising features of holography that have to be standardized, e. g.

— high-resolution colour holography, which is particularly important to record objects in natural colours [7],
— mutiplexed holographic recording in order to store the information of all sides of the original object as well as different illumination directions within one hologram,
— optical filtering and correlation to recognize handwritings [8].

A standard has now been reached that opens the chance to introduce holography as a regular tool for epigraphists and archaeologists and other scientists working in the field of humanities.

Furthermore, in this way the information content of those valuable cultural objects is secured. Museums and collections can complete exhibitions where certain originals are not at their disposal and specialists involved in such studies can obtain hologram copies in order to complete their individual archives.

References

1. C.B.F Walker: Cuneiform, Berkeley/London (1987)
2. B. Kienast: The Old-Assyrian Texts of the Oriental Seminar of the University of Heidelberg (in German) **24** (1960)
3. D. Vukicevic, G. von Bally and W. Sommerfeld: "Holo-Paleo-Graphy" holography in protection, preservation and evaluation of ancient artifacts with cuneiform inscriptions, Optics Within Life Sciences (OWLS), Elsevier, Amsterdam **2** (1993) 153–157
4. G. von Bally, F. Dreesen, V. Markov, A. Roshop, E. de Haller: Registration of color holograms on PFG-03-C monolayers, Journal of Technical Physics **21** (1995) 76–80
5. N. Demoli, G. Wernicke, H. Gruber, U. Dahms: Holographic techniques application in analyzing cuneiform inscriptions, Journal of Modern Optics **42** (1995) 191–195
6. G. von Bally, D. Vukicevic, N. Demoli, H. Bjelkhagen, G. Wernicke, U. Dahms, H. Gruber, and W. Sommerfeld: Holography and holographic pattern recognition for preservation and evaluation of cultural-historic sources, Naturwissenschaften **81** (1994) 563–573
7. G. von Bally, F. Dreesen, A. Roshop, E. de Haller, G. Wernicke, N. Demoli, U. Dahms, H. Gruber, and W. Sommerfeld: Holographic methods in cultural heritage preservation and evaluation, Optics Within Life Sciences (OWLS), Elsevier, Amsterdam **3** (1994) 297–300
8. N. Demoli: Optical correlation methods in epigraphy, Optics Within Life Sciences (OWLS), Springer, Heidelberg **4** (1997) (this issue)

This work is supported by the Ministry of Education, Research, Science and Technology of the Federal Republic of Germany within the research Program on "New Technologies in the Humanities" (NTG) (03VB9MU18).

Investigation of Decay Mechanisms in Historical Artwork via TV Holography

G. Gülker

Carl von Ossietzky University Oldenburg, FB Physik, Postfach 2503, D-26111 Oldenburg, Germany
E-mail: guelker@uwa.physik.uni-oldenburg.de

Abstract. The preservation of works of art calls for the investigation of underlying corrosive processes. TV-holography has proved to be a powerful tool for this purpose. A review will be given of various applications in diagnostics of works of art.

1 Introduction

The deterioration of historical valuable works of art is of major concern, as these objects represent part of our cultural heritage. Their preservation is thus a priority task in order to maintain these tokens for the next generations of mankind.

The causes of deterioration are consequences of century-long climatic impacts aggravated by man-made pollution of recent years. Effective preservation calls for the investigation of underlying corrosive processes, which are not yet fully understood [1]. To take care of the delicate objects and for an early diagnostic warning, non-destructive and highly sensitive techniques are most favourable. Thus, optical inspection with interferometric sensitivity is extremely well suited for this task. However, investigations under adverse conditions outside the laboratory at a monument site must meet special requirements. In the last couple of years, TV-holography or electronic speckle pattern interferometry (ESPI) has proved to be the favourite candidate for this purpose.

Introduced by Butters and Leendertz in the early 1970s [2], TV-holography is basically a holographic method. Instead of using photographic films, the recording of the interference patterns is performed by a video camera, avoiding the troublesome chemical developing process. It has been known for a few years that diagnostics and conservation of works of art can greatly benefit from TV-holography, too. However, the method is far removed from a widespread utilization in the field of conservation. Consequently, there are only few research groups using TV-holography intensively for preservation purposes. The method offers almost the same variety in different modifications as conventional holography. Depending on the specific realization of the experimental setup, the method reveals a variety of useful information about the historical artefacts under investigation. This paper reviews the state of the art in the application of TV-holography to diagnostics of works of art.

2 Deformation Mapping

One of the most important applications of TV-holography is the study of minute surface displacements that are connected with omnipresent physical and chemical impacts. The monitoring of tiny deformations due to e.g. climatic changes or corrosion can help to obtain insight into decay processes, identify object areas of special concern, yield material properties or serve to control the success of conservational remedies.

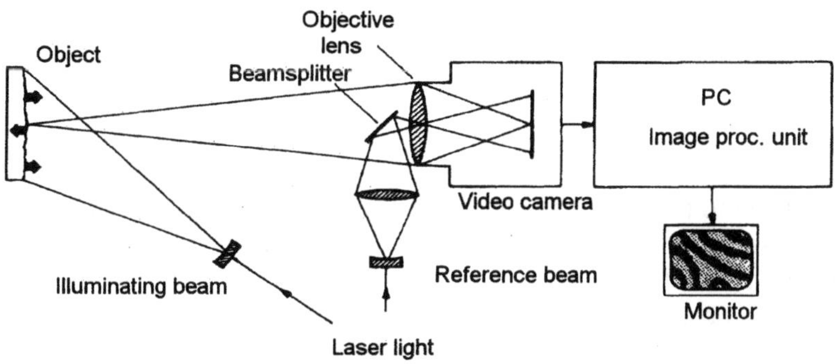

Fig. 1. Scheme of a typical TV-holography setup for monitoring static deformations.

In Fig. 1 the schematic setup of a typical TV-holography system is presented that is able to measure static deformations on a micrometer scale. Since the method is comprehensively described in literature, e.g. in [3], only the principle of operation is outlined here. The light of a laser illuminates the surface of the object under investigation. Meanwhile, small and compact diode lasers in the near infrared wavelength region are utilized, in many cases with output powers in the order of 100 mW. This is sufficient to illuminate areas up to about a quarter of a square meter. Due to the coherence of the laser light, a speckled image of the surface is formed on the target of a video camera, typically a small and handy CCD camera. The field of view can be adjusted by a zoom lens or exchangeable optics. To obtain interferometric sensitivity, a part of the illuminating wave is divided by a beamsplitter and superimposed on the objects image. Hence, an image plane hologram is created on the target of the CCD-camera.

Instead of storing the superimposed light fields on photographic media as in traditional holography, the interference patterns are captured electronically. The holograhpic information can be digitized and processed with a PC-based image processing system. In standard applications, holograms of different states of the object under investigation are compared by digital addition or subtraction. In the case of object deformation this results in an image covered with dark and bright fringes connecting locations on the surface of equal amount of surface displacement. The results are displayed on the monitor and can be stored on a video tape recorder or in the computer. The method operates at the speed of the video system (i.e. 25 to 30 Hz) and is thus well suited to monitor the time development of a displacement field in video real-time. Like in conventional holography, the sensitivity vector of the arrangement bisects the illumination and observation angle. Thus, the setup in Fig. 1 is sensitive mainly to out-of-plane displacements.

To demonstrate the performance of the system and the type of information that can be obtained, some examples will be presented. At this point it is worth mentioning the important work of four groups in Europe using TV-holography in the investigation of works of art: these are the groups of Paoletti and Lucia in Italy; Markov in the Ukraine; and Boone in Belgium. Investigations on medieval frescos [4], ancient violins [5], or museum objects [6] are only a few illustrations of their activities. However, since all of these groups contribute to this volume describing in detail part of their work, examples described in this paper will be restricted to the author's own investigations to avoid redundancy.

To begin with, the behaviour of a historical natural stone probe (Fränkischer Sandstone) on simulated humidity attack was investigated. Due to its damaged state the stone probe, which originally was part of the 'Würzburger Residenz' in southern Germany, had to be replaced by a replica some years ago. It thus represents a superior test object to study material properties as for example hygroscopic dilatation.

In Fig. 2a the test object is shown with the optical head of our TV-holography system fixed to the stone probe. In most on-site measurements we establish a direct connection between the head of the system and the object under investigation. This minimizes relative movements between measuring head and object and yields very stable setups. With such coupling, continous interferometric measurements were performed over long time periods, e.g. several weeks [7]. If only short time investigations are desired, measurements can be performed from a stable tripod without any mechanical coupling [4]. In some cases, the modified method TV-shearography [8] can be used, which is a common path interferometer and thus very insensitive to external disturbances [9]. However, it has the great disadvantage that only the slopes of occuring deformation can be detected.

For the preparation of the simulation experiment, a small hole was drilled into the material. The area of investigation, which extends about 3.5x3.5 cm^2, is shown in Fig. 2b in a close-up photograph. In the beginning of the experiment, the hole was filled partly by a very small amount of water. The micromechanical response of the stone was monitored by TV-holography. In Fig. 2c a typical primary resultant image of the method is shown, produced by subtracting an image of the unloaded stone from an image grabbed about one hour after wettening the stone. Concentric fringes around the bore hole are visible, connecting surface areas of equal amount of deformation. When the fringe spacing is known, i.e. the difference in deformation between two adjacent fringes, which typically is in the order of half the wavelength of the laser light (i.e. in the order of 0.5 μm), and after data processing performed by the computer, the appropriate deformation occuring at each surface point can be extracted automatically. In Fig. 2d a 3-D representation is shown indicating a nearly symmetric deformation around the bore hole due to hygroscopic dilatation. It has to be considered that the maximum deformation value occuring in this Figure is only about 2.5 μm.

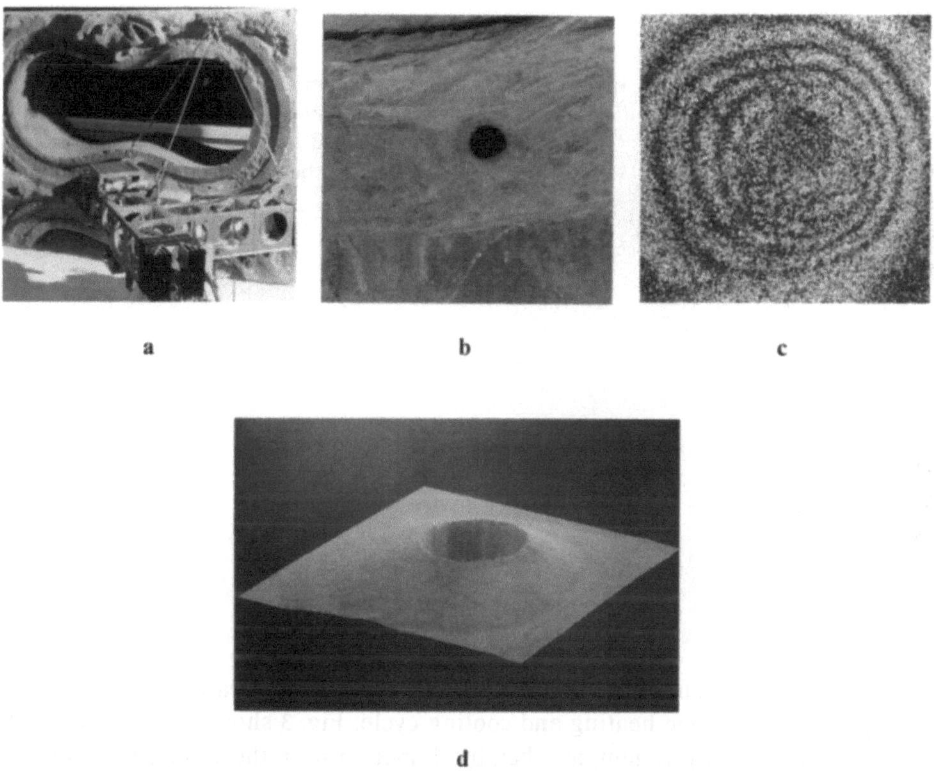

Fig. 2. Hygroscopic dilation measurement. (a) Optical head of TV-holography system fixed to test object (Fränkischer sandstone from 'Würzburger Residenz'), (b) area of investigation (3.5x3.5 cm^2) with borehole, (c)resultant fringe image showing deformation of stone area around the borehole due to wettening and (d) 3-D representation of the occured deformation.

It was shown in this measurement that material properties like hygroscopic behaviour can be studied by TV-holography. In comparison to pointwise measuring methods, a two-dimensional deformation field is obtained. Since the onset of delapidation often shows up in surface areas with irregular behaviour, as for example inhomogenious expansion in comparable experiments, this can be easily identified by TV-holography.

In a further example, deformation measurements on a baroque mural from the 17th century inside a church located on the small German island of Föhr were performed. The vaults of the inner part of the church are covered with paintings which show pronounced deterioration. In this case it was expected that many of the deterioration processes could have been initiated by the periodical heating of the church during winter time. Volume increase during crystallization of salts present in the wall and plaster was suspected to be responsible.

Fig. 3. Deformation monitoring on baroque mural. Optical head of the TV-holography system fixed to the wall to minimize movements between optical head and wall.

To prove this assumption, deformation measurements on a part of the mural were performed for one heating and cooling cycle. Fig. 3 shows the optical head of the TV-holography system, at a height of about 6 m in the vaults, fixed to the wall to minimize movements between optical head and wall. Serveral hours after fixation of the optical head, the heating system was activated and occuring deformations were recorded. After a while, a very peculiar deformation in a part of the test area could be observed. In Fig. 4 the deformations during the measurement period are shown in a 3-D representation.

A macroscopical crack transversing the test area (about 5x5 cm^2 in extend) is clearly visible in the center of the 3-D deformation map as an area of deep depression. This was not unexpected since the crack was visible with the naked eye. However, the right side of the test area showed a nearly bump-shaped deformation with pronounced deformation gradients, which seemed to be caused by a relatively localized pressure from beneath the surface. This supported the assumption that crystallization of salt could be responsible. Thus, for further verification a plaster sample of this region was removed and examined by scanning electron microsopy immediately after the peculiar deformation was detected. Indeed, a high amount of salt crystals, partly of large size, were found in this specimen, which is shown in the micrograph of Fig. 5. It is not hard to imagine that frequent cycles of cystallization and dissolution finally could lead to detachment of plaster parts. As a consequence, a modification of the heating system is in the blueprint stage to ensure only very moderate climatic changes inside the church.

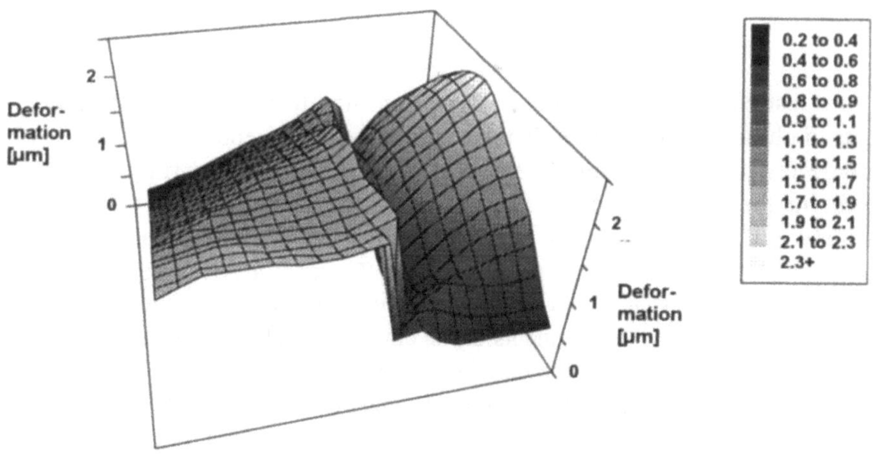

Fig. 4. 3-D representation of salt-induced wall deformation during heating phase.

Fig. 5. Micrograph of plaster sample taken from bump-shaped deformation zone immediately after detecting identifies salt crystalls of partly big size.

In a last example concerning deformation monitoring, investigation of fragments of chinese Qin Shihuang's terracotta warriors is reported. In March 1974, Chinese farmers at Lintong, 35 km east of Xian, discovered fragments of clay figures in the vincinity of the mausoleum of the first Emperor Qin Shihuang (249-210 B.C.). Three months later the excavation of the most spectacular archaeological find of this century began: the Terracotta Army of the Emperor, consisting of more than 7000 life-sized clay soldiers, horses, wooden wagons and bronze weapons. The fragments of the clay figures show immense sensitivity to humidity changes, and since the excavation led to a desiccation of the figures, the remnants of the paint layers were extremely endangered to fall off. In order to develop suitable methods of conservation, the excavation work on the figures was stopped several years ago. A joint research program between the Ministry of Cultural Properties of the Province Shaanxi and several European institutes under guidance of the Bavarian State Conservation Office in Munich was started.

In the context of this cooperation, TV-holographic measurements were performed by our group. The aim of this investigation was to monitor deformations of clay fragments and the fragile ground layer pieces due to humidity changes, and to estimate the influence and suitability of several conservation agents and procedures. To perform humidity variations, the objects under investigation were inserted into a small computer controlled climatic chamber where they were exposed to well defined ambient humidity. An air stream of definable temperature and relative humidity generated by means of two thermostats was guided through the small chamber to control the climatic condition. Sensors for temperature and humidity were inserted to measure the climatic situation inside the box.

Fig. 6. Fragment of chinese terracotta warrior. Tested area of about 2x1.5 cm^2 marked by a rectangle (Photo by C. Thieme, Bavarian State Conservation Office, Munich, Germany).

In a series of experiments, deformation measurements were performed on a couple of terracotta fragments partly covered with pigmented layers and on several pieces of isolated ground layers. The samples were subjected to different conservation treatments. In Fig. 6 one of the more than 2000 year-old terracotta fragments is shown. It is almost totally covered by a dark brownish ground layer and only in the mid left of the specimen did a small remnant of light brown colour remain on the surface, which is not recognizable in the photograph in Fig. 6. The specimen was treated with a 'Bologna Cocktail', a very specific mixture of diverse treatments. The tested area of about 2x1.5 cm^2 is marked by a rectangle.

In the beginning of the experiment the specimen was inserted in the climatic chamber at a relative humidity of about 61%. When loading of the chamber was finished, the humidity was increased. Surprisingly, the treated colour layer instantaneously reacted to humidity change. A bump-shaped deformation zone was detected which coincides exactly with the size of the light brown-coloured patch, while the rest of the sample remained undeformed. This was not expected in advance. Fig. 7 shows the resulting fringe patterns at this stage: an almost concentric fringe system at the location of the treated colour layer piece can be recognized. Fringes represent the deformation that occurred during a humidity raise of only about 0.2% r.H. This shows the extreme sensitivity of the layer to humidity changes. A further very astonishing point is the direction of the deformation: although the humidity was increased, the fringes are characterizing a displacement away from the observer, that means a bump-shaped deformation into the surface plane. The reason for this is still a matter of speculation.

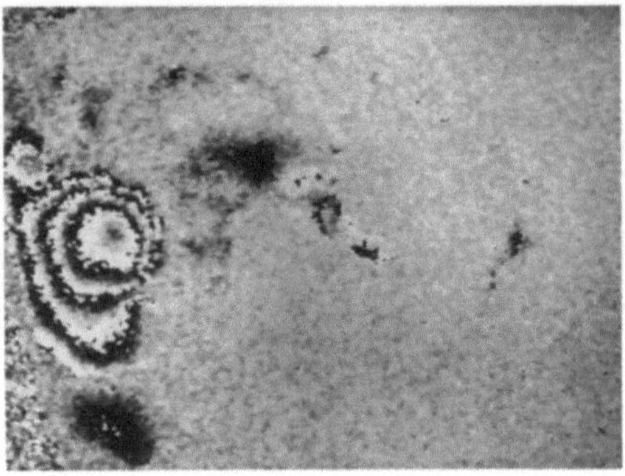

Fig. 7. Nearly concentric fringe pattern indicates deformation due to humidity change of only 0.2% r.h! Deformation area coincides exactly with size of coloured relics. Investigated area of about 2x1.5 cm^2 is marked by a rectangle.

To perform a humidity cycle and to verify deformation in the opposite direction, humidity increase was stopped at about 95% and was decreased again, down to a final value of about 58%. It was very interesting that, in fact, the deformation direction inverted and that this turned up with a small delay: the reverse displacement started at about 94% and about 8 minutes after decreasing the humidity (at this point of the experiment, humidity decrease of 1% took about 8 minutes; later on, decrease was much faster). The occurring deformations were evaluated quantitatively over the whole decreasing period. It was found that a maximum deformation, which is equal to the height of the bump, of about 280 μm occurred. If a linear behaviour is assumed, a negative deformation rate of about 8 μm/%r.H. was obtained.

Furthermore, the examined region seemed to be very stable. A humidity decrease down to about 60% did not lead to any detectable kind of damages, e.g. cracks or detachments etc. However, just below 60%, deformation in this area became more asymmetric. At about 58% r.H. at least one deformation fringe showed a sharp bending, probably indication of the onset of dilapidation. The experiment was stopped immediately at this point in order to preserve the sample.

It was shown in this example that TV-holography in combination with a climatic chamber was very well suited for studying various severe dilapidation processes acting on these vulnerable historical materials. Questions concerning the influence and suitability of conservation treatments could at least partly be answered.

3 Mural Examination by Vibration Analysis

So far, examples have been given where TV-holography was used to measure static object micro deformations. However, as stated in the introduction, the method is not limited to deformation measurements. TV-holography has been known for a long time to be a powerful instrument for analyzing object vibrations, too. Several examples are given in the literature where this method was used successfully to investigate e.g. vibrating car doors or electronic devices [10][11]. In this chapter another rather new application field of TV-holography, the detection of plaster detachments, is presented.

Changing environmental conditions lead to accelerated deterioration of historically valuable murals and require restoration steps in ever shorter time intervals. Driven by fluctuations of humidity and temperature, especially crystallization and dissolution cycles of salt in the wall cause partial detachment of painted plaster layers from the supporting wall [12]. For the preservation and restoration of such delicate artwork it is important to identify the debonded regions early and to map them quantitatively to obtain the current state of the preservation and to determine deterioration development. Such diagnoses are the basis for planning and conducting any restoration.

A commonly used technique to localize detached areas is the so-called percussion method: the restorer knocks slightly with his finger against the test section and estimates the quality of the bond from the acoustical response. Obviously, the method has several disadvantages: the knocking procedure is very cumbersome and time consuming, especially if repeated exploration of large wall sections is required. Furthermore, the method is intrusive because the finger contact may affect the delicate layer of paint and expensive scaffolding is usually necessary. Finally, the results may differ depending on the individual experience of the restorer. The approach for a new non-invasive method for remote monitoring of detached plaster areas utilizes acoustical excitation of the wall combined with optical registration of its vibrational response by TV-holography.

Physically, a loose plaster layer in front of a rigid wall forms an oscillatory system comprised of the plaster mass and the compliance of the air gap between plaster and wall. This system can be excited remotely to minute vibrations by directing sound waves against the wall. Since excitation is most effective at resonance, the acoustical driving frequency is swept through the resonance which varies depending on the condition of the particular wall-plaster system. The protection of the delicate artwork as well as an acceptance of the method in practice calls for minimum sound pressure requiring very small vibration amplitudes even in the resonant case.

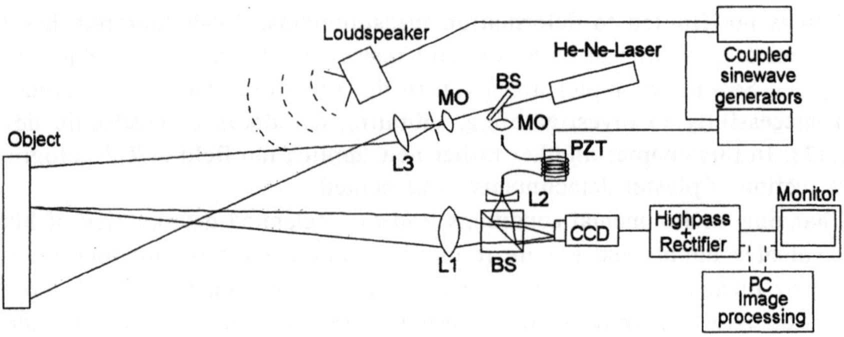

Fig. 8. Experimental setup for vibration monitoring. (MO: microscope objective, PZT: piezoelectric cylinder, L: lens, BS: beam splitter)

In Fig. 8 the basic configuration of the method is outlined. Again, the optical part consists of a TV-holography setup. However, its design is slightly modified to perform time-averaging vibration measurement. Vibration measurement by TV-holography is reprorted in detail e.g in [10] and thus, only a brief description of the method follows. The object under investigation, in these applications a section of a plastered wall, is excited by sound waves generated by a loudspeaker and is illuminated by the light of a laser. The backscattered light is collected by the objective lens and thus fed into the interferometer. A distinction from the former setup is that no computer or image processing unit is necessary. Instead of image subtraction etc., image plane holograms are time averaged for each video cycle, in principal very similar to time averaging holography. The camera signal is then rectified and high-pass filtered in analog-electronic circuits and finally displayed on a monitor.

Standard time-averaging vibration measurements yield a resolution in the order of $\lambda/5$, where λ represents the wavelength of the laser light, presently 633 nm. It has been demonstrated that high sensitivity of some tenth of a nm can be achieved by periodically changing the path length of the reference beam at an appropriate frequency [11]. For this purpose, the reference beam in Fig. 8 is guided by a mono-mode optical fiber, which is partly coiled around a piezoelectric cylinder PZT, a device that changes its diameter when applying a voltage. The necessary path length modulation of the reference beam is achieved by stretching the fiber appropriately. As a result of this method, vibrating regions representing loose plaster parts in the investigated wall section show a time-varying brightness on the monitor and even very small amplitudes can be identified due to the flickering in the life images on the screen. With our laboratory system a lower resolution limit of about 2 nm vibration amplitude was achieved.

A first test of this method under realistic conditions was performed at the medieval mural of the apse in the church of Eilsum, in northwest Germany. The mural has been rapidly dilapidating since it was uncovered in 1970 because of extremely high amounts of salt and humidity trapped inside the walls. Furthermore, microbiological attack on and behind the painted plaster was found. Fig. 9 shows part of the apse carrying the mural originating from the 13th century. Humidity and temperature detectors can be recognized in the front and at the top of the left side in the Figure. The very poor preservation state of the mural is clearly visible.

The test region investigated by the new method has a size of about 25x40 cm^2 and is located above the window on the west side about 6 m high at a clearly vaulted part of the ceiling. The optical setup was placed on the bottom of the apse and pointed towards the test region. The loudspeaker for the acoustical excitation was mounted at the chorus at a distance of about 5 m from the test region. The frequencies of the acoustical excitation and reference beam modulation were swept from 100 Hz to 5 kHz to find the resonances of the debonded areas. The resulting TV frames were recorded on a videotape for later evaluation and for a possible comparison with future investigations. The time needed for the inspection of the whole area was only about three minutes.

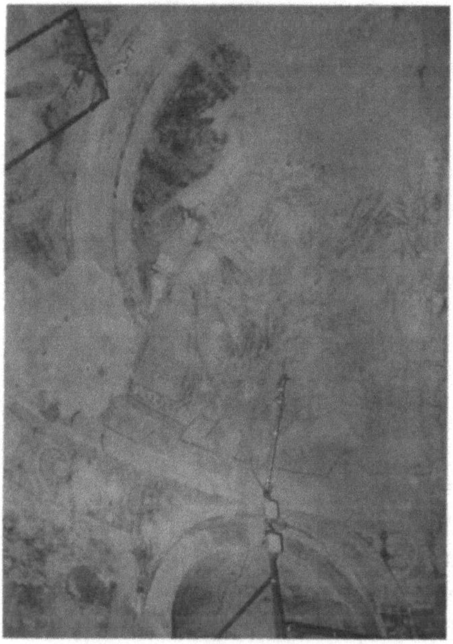

Fig. 9. The 13th century apse in the church of Eilsum, Ostfriesland, north-west Germany, site of measurements.

The first out-door application at this historical site confirmed that the new method is very well suited for the monitoring of detached plaster zones. Although the quality of the recordings was not as good as the laboratory results (mainly due to air disturbances), vibrating plaster areas could be identified easily by directly observing the flickering brightness in the live images. In Fig. 10, the resulting map of detected loose areas within the test region is presented. The hollows that were visible at different excitation frequencies are marked by different hatchings. The biggest one on top of the left side has a diameter of about 10 cm and is resonant at the lowest observed resonance frequency of 590 Hz. More detachments of smaller extensions were found at higher resonance frequencies. In spite of the partly small sizes of the vibrating regions, the resonant frequencies were considerably lower than in comparable laboratory measurements at a test wall. Thus, it can be concluded that the strength, cohesion and elasticity of the historical mortar are lower. The sound pressure level was measured about 85 dB at a height of two meters in front of the wall below the tested region, which corresponds to very loud traffic noise. The induced vibration amplitudes are about 20 nm. In comparision, organ music may exceed sound pressure levels of more than 90 dB. Therefore, damages due to the applied sound pressure are not expected. Further tests are presently performed to verify this assumption.

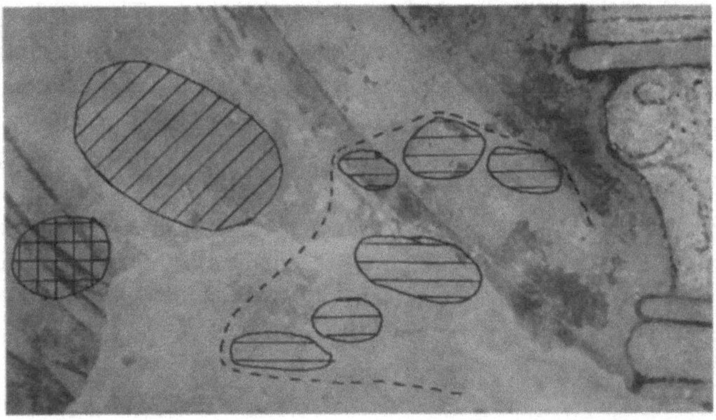

Fig. 10. Results of optical vibration test: Photograph of the investigated region (about 25x40 cm^2) located on the ceiling vault at 6 m height. Hatched regions mark partially detached plaster found by acoustically excited minute vibrations at different frequencies. //////: 590 Hz; ####: 1060 Hz; ═══: 1.3 - 1.8 kHz.

4 Surface Inspection by Decorrelation

All previously described modifications of the TV-holography technique exploit object information contained in the scattered light speckle field to identify object microdisplacements. In addition, TV-holography subtraction fringes also contain information about any decorrelation of the two governing speckle patterns. Such decorrelation effects impairing TV-holgraphy measurements are well-known [13]. They can be used, as we will see, beneficially for a quantitative analysis of the process responsible for the decorrelation.

Many deterioration processes show up in the microcosm on the surface of the stone. Condensation and evaporation involve changes in the distribution of soluble materials. Aggresive pollutants induce chemical reactions, airborne matter is deposited, and microorganisms grow cultures. Decorrelation studies are therefore an ideal means to monitor surface processes.

To understand the principle of the modified technique [14] it is useful at this point to recall the formation process of TV-holography fringes. Deformation produces a shift in the phase difference between object speckle pattern and reference wave. The resulting image will be very similar to the original image when the change is close to an integer multiple of 2π. It will be a more or less different speckle pattern for intermediate phase values. The subtraction image thus consists of dark areas where the very similar speckle patterns cancel each other out, and of light areas, in which the subtraction process produces just a new speckle pattern. When, in addition, the scattered object light field changes, the dark areas are also gradually covered by speckle thus reducing the visibility of the fringes. Thus, if the mean intensity of a small area in the fringe system is taken continuously while the fringes are shifted over at least one fringe period, its modulation is proportional to the correlation of the underlying speckle patterns. To perform the required shifting of fringes without any principal variation of a typical setup and without additional equipment, the phase-shift method [15] is used. Applying this method, a modulation index for each averaged subregion can be calculated from three or four fringed images by introducing a small appropriate variation in the phase of the reference beam between subsequent frames. The computation of the modulation index field is performed quite rapidly in the image processing board of the personal computer. The modulation is then attributed to one of 256 grey levels on the monitor. As a result, the intensities in the final image are proportional to the amount of correlation of the two compared light fields. Bright regions represent high correlation, low intensities correspond to areas of reduced correlation. In the absence of decorrelation effects due to external influences, as for example rigid body motion of the object, a decrease in intensity is directly related to the amount by which the surface microstructure changes.

The method was used to compare water-induced microstructure changes in samples of different types of natural stone. Fig. 11a shows the test specimen, a compound of two sandstones (at the top a Saaler, in the middle a Schönbucher sandstone) and a steel plate (at the bottom). This sample of about 5x5 cm^2 surface area was exposed to a cyclic change in humidity of the ambient air from 27% to 90% relative humidity and back to 27%. Figs. 11b-d show the correlation maps at three different instants during the experiment. In Fig. 11b, which was taken at 59% r.h., the decreasing brightness in the regions of the stone samples indicates decorrelations. Changes in the Schönbucher are stronger than in the Saaler sandstone. The steel surface shows no decorrelation. Fig. 11c presents the situation at 90% r.h. While the steel plate remains nearly unchanged, the reduced intensity in the two stone samples indicates considerable changes in the microstructure at the surfaces. Reactions in the Schönbucher sandstone are still more pronounced than in the Saaler sandstone. This can be explained by microscopic and mineralogical data on the stones: They differ in microporosity and roughness. Furthermore, the Schönbucher sandstone is equipped with more

9. G. Gülker, H. Helmers, K.D. Hinsch, P. Meinlschmidt, and K. Wolff, J. Opt. and Lasers in Eng, 24 183-213 (1996).
10. O.J. Løkberg, J. Acoust. Soc. Am. 75, 1783-1791 (1984).
11. K. Høgmoen and O.J. Løkberg, Appl. Opt. 16.7, 1869-1875 (1977).
12. P. Mora, L. Mora, and P. Phillipot, *Conservation of Wall Paintings* (Butterworths, London, 1984).
13. J.T. Malmo and O.J. Løkberg, in: *Laser Interferometry: Quantitative Analysis of Interferograms*, Proc. SPIE 1162, 270-278, (1990).
14. G. Gülker and K.D. Hinsch, J. Opt. and Lasers in Eng 26, 165-178 (1997).
15. K. Creath, in: Progress in Optics XXVI, 351-393, Elsevier, New York, (1988).

Application of Holographic Interferometry to Museum Objects

Pierre M. Boone*, Vladimir B. Markov
Nikolay M. Burykin, Vadim V. Ovsyannikov

*University of Gent, Department of Applied Mechanics
Laboratory Soete
Sint-Pietersnieuwstraat 41, B-9000, Gent, Belgium
tel. +32-9-2643242 fax +32-9-2237326 e-mail Pierre.Boone@rug.ac.be

Institute of Applied Optics
National Academy of Sciences - Centro Internacional de Fisica
Kudryavskaya 10-G, 254053, Kiev, Ukraine - Apartado Aereo 49490, Bogota, Colombia
tel. +380-44-2122158 tel. +57-1-2692789
fax +380-44-2124812 fax +57-1-2682366

Abstract. In this paper some experimental results related to practical applications of holographic non-destructive techniques for the inspection of the condition of museum items are presented, especially those used in the detection of deformation, stress concentration and defect localization. These results are obtained using the basic principles of traditional double-exposure holographic interferometry, as well as with its electronic variant with image processing. In the latter case, the accuracy in displacement measurements was studied on a specific test-object. Two museum items were selected for their condition inspection: a XVIIIth century icon on wooden panel and a precolumbian terracotta anthropomorphic. The described technique can give quite accurate metrological information, but is also rather promising for rapid qualitative analysis of the object condition before and after restoration, as well as during its storage.

Keywords: holographic interferometry, non-destructive testing, metrology, art work diagnostic.
Abbreviated title: Video-holographic interferometry applied to museum objects

Although the results obtained are convincing, conventional **HNDT**-methods using photographic type recording materials are not very useful due to the lengthy processing and the long time interval between recording and analysis. Taking into account the number of the objects to be studied and the necessity of analysis of the obtained experimental data, a general use does not seem to be very realistic. That is why electronic based systems with image processing seem more suited for those kinds of applications[14,16].

2.2 Holographic video technique

The experimental system, developed for museum applications, consists of optical and electronic components, as well as specially elaborated software. In the optical part, an ion argon laser LEXEL Model 95 was used as a light source. It was operating in a single frequency regime with an output power of about 200 mW at λ = 514 nm (it was also possible to use a 50 mW single frequency He-Ne laser LGN-220 to control small size objects or small local areas). The laser beam was split with a semi reflecting mirror into a reference and an object beam. The latter was expanded up to 80 cm diameter and its intensity at the object's plane did not exceed $3{\cdot}10^{-2}$ mW/cm^2.

The electronic part of the system is for "holographic" image acquisition through a CCD camera. It was also used to control the position of the mirror attached to a piezoelectric transducer (**PZT**). This mirror was for phase-stepping of the interference pattern. The images that correspond, respectively, to *original (non-disturbed)* and *disturbed* states of the object, were taken by a standard SONY CCD camera and digitization was done at 512 x 512 pixels with 256 grey levels, supported by an image acquisition and processing board DT-2856 with specially developed software ***IIP*** (Interferometric Image Processing). For more details on this, see e.g. ref (18).

Preliminary experiments with a membrane as test-object were used to check the system operation. The 50 mm diameter elastic diaphragm was used as a membrane. It was fixed to the holder along its contour, and a piezoelectric transducer **PZT/O** was attached to the center of this diaphragm (Fig 2.) The **PZT/O** shifting constant was ~ $2.5{\cdot}10^{-2}$ μ/V. Then the profile of the membrane was changed by applying a suitable voltage. In these experiments "zero" level displacement corresponds to the boundary area of the diaphragm, where the membrane is rigidly attached to the holder. An atypical picture of the interferogram is shown in Fig. 3. The concentric interference fringes confirm the uniform and symmetric bending of the membrane.

To estimate the precision and accuracy in displacement measurements, a set of experiments has been performed with the membrane. Typical variations of displacement distribution over the diaphragm surface are presented in Fig. 4 a,b. It follows from these data that the sensitivity of the system is about 0.15 μ, with voltage applied to **PZT/O** less than 5 V.

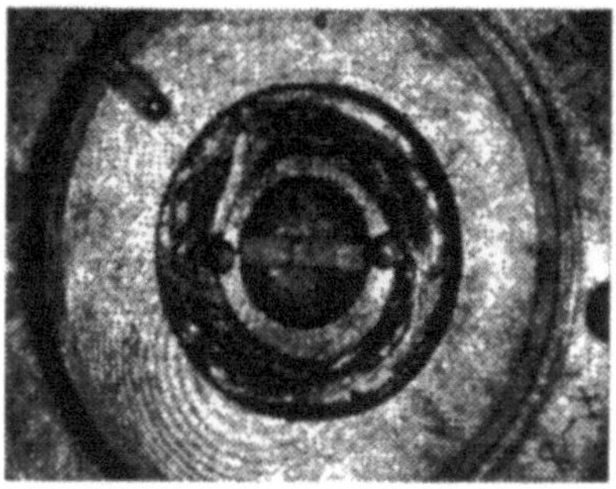

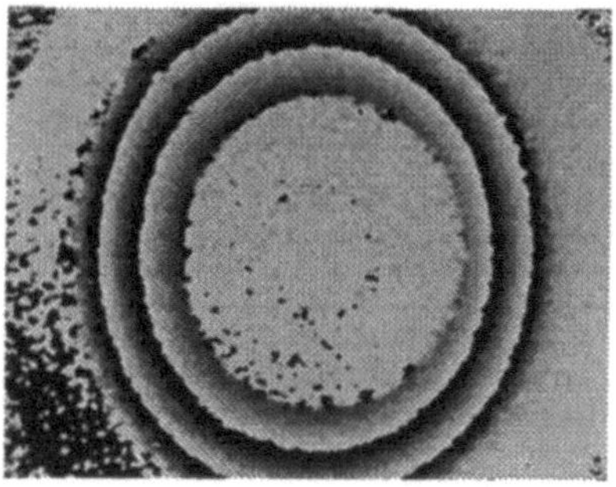

Fig. 2. The membrane test-object, as seen through CCD camera.

Fig.3. Interferogram on membrane.

Evidently, increasing the voltage leads to membrane bending superimposed on stretching, and the value of this bending can be determined from interferometric measurements. Fig. 4 b shows the membrane cross-section after a bending force has been applied. These results are obtained from the processed interference pattern of Fig. 3 b.

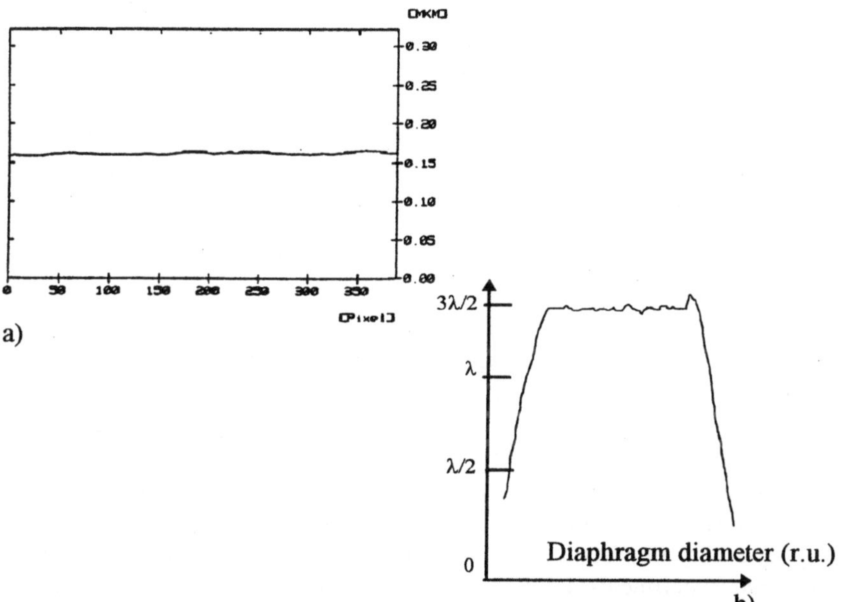

Fig. 4 a,b. The result of holographic interferogram analysis for membrane bending with different level of voltage applied to **PZT/O**: a $\leq$ 5 V; b $\approx$ 38 V.

The relation between measured and induced displacements are shown in Fig. 5.

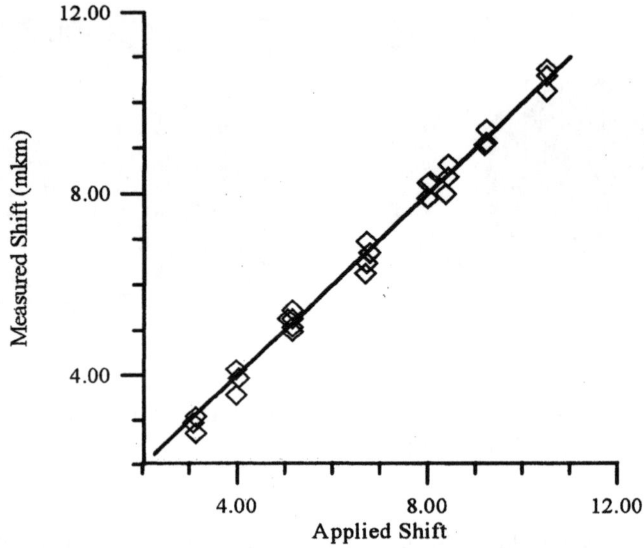

Fig. 5. Normalized dependence of holographicaly measured shift of diaphragm with initiated one.

3. EXPERIMENTAL RESULTS

The second test object was a small terracotta figurine (height approximately 65 mm) of the Columbian Terradientro culture, probably about 6th century BC. Only part of it was tested, about 50 mm high (Fig. 6). Visually, one is able to detect a crack starting from the outward corner of the left eye and stopping at the shoulder. However, when looking at the interference fringe pattern in Fig. 7 (mechanical loading, details see later), one sees that the discontinuity goes a lot further, and therefore the defect is not as innoxious as it looks. Fig. 8a shows the phase unwrapped version of the fringe pattern, Fig. 8b the 3D-representation of the displacement parallel to the observation direction, i.e. almost normal to the surface visible in Fig. 6.

With respect to classical "photographical" holographic interferometry, one has the advantage that it is possible to "see" what is happening during loading, as one can observe the fringes in "real" time (in fact, at videoframe rate). In the case of this pattern, loading was effectuated by gently pushing on the head of the statuette with one finger, while the other hand of the experimentator held the legs in place. Firing of the laser was done by foot work.

Finally, and very important from the point of view of possibilities of wider application, some experiments were performed on an XVIIIth century Ukrainian icon. This icon is very typical for objects with multi-layer structure. It has a wooden support coated with several different ground (priming) sub-layers, and has been gilded over the entire panel area. Then, this gold surface was partially painted with oil-paint.

Figure 6

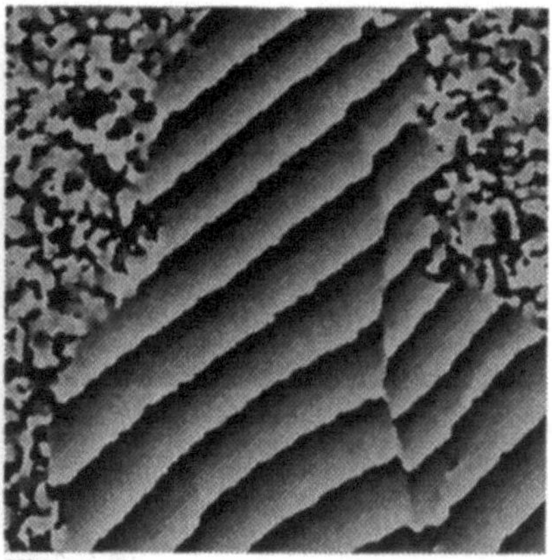
Figure 7

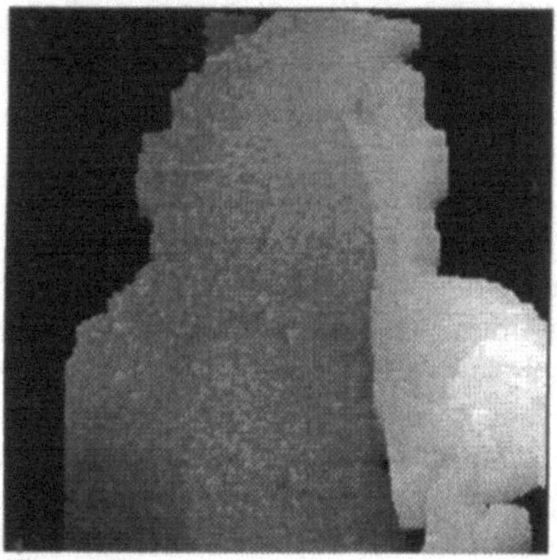
Figure 8a

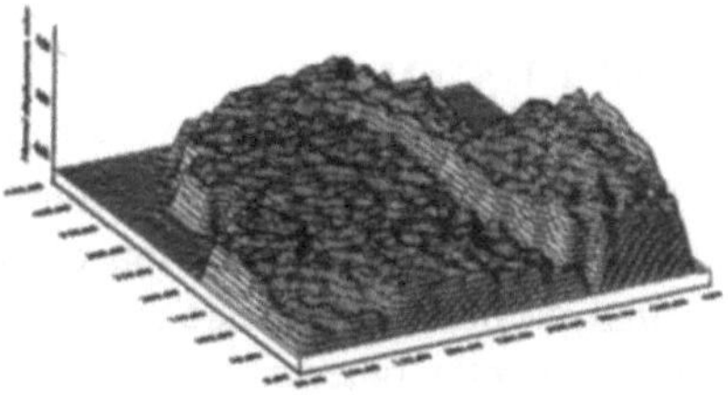
Figure 8b. 3D representation of displacement field.

The problem of such multi-layer structures is normally connected with formation of local disbonds between priming layers and the wooden panel on the one hand, and between gilded surface and sub-layers on the other. This can be a result of humidity and/or temperature variation, mechanical instability, etc. All these can give rise to the cracking of the oil-paint and its surface, which can finally damage the object.

We were faced with the problems of high light reflectivity by the gold surface. In this particular situation, it was impossible to use any kind of coating powder to prevent bright light spots reflected by the metal (the technique used in engineering is to spray talcum or calcium carbonate powder gently over the object surface). So, some measures had to be taken to guarantee the proper image acquisition; details will be given in a further publication. Fig. 9 (a) presents the view of the icon's part as seen through the CCD camera. The IR-lamp used was a 250 W placed at about 1,5 m distance from the back of the specimen, resulting in a temperature rise of about 0.5°C. It resulted in the interference pattern shown in Fig. 9.

The observed non-uniformity in the fringes distribution can be attributed to local disbond and classified in this particular case as detachments (Fig. 9 c) between priming and gold layers. So, increasing of the icon surface temperature has resulted in de-bonding effects.

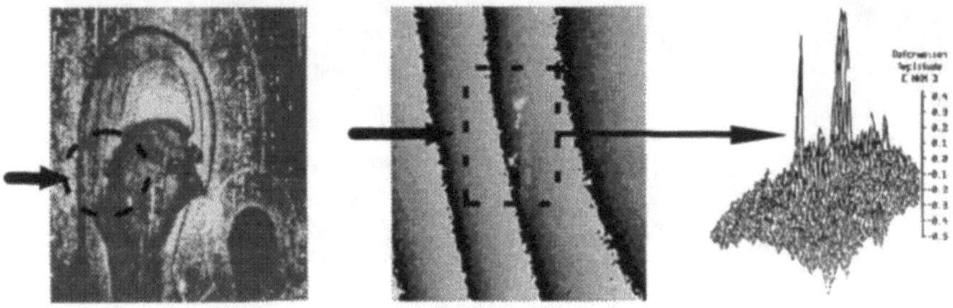

a). b). c).

Fig. 9 a,b,c. Fragment of a XVIIIth c. Icon as seen through CCD camera (a), its interferogram with defect localization (b) and pseudo-3D view disbond area (c).

4 CONCLUSIONS

Video aided holographic interferometry is a very promising tool and technique for the non-destructive evaluation of condition, as well as defect localization and stress analysis of artwork made of various materials, such as oil paintings on wooden panels, terra-cotta objects, etc. The most important advantages of the developed system are: its relatively high speed of operation; high light sensitivity; it is a full-field method that requires only a very small amount of loading, normally less than classical variations of environmental changes in usual museum conditions. It is also demonstrated here that it is possible to obtain quantitative measurements with this method, when supported with specially developed software.

5 ACKNOWLEDGMENTS

This work was done partially at Gent University (Belgium) where one of the authors (VBM) spent three months as a visiting professor. The financial and technical support from Gent University and the Belgian National Science Foundation through grant 2.004486 are gratefully acknowledged.

Part of the research activity was performed in frame of the scientific program of the International Institute of Applied Optics, an initiative of the National Academy of Sciences of Ukraine, and the Centro Internacional de Fisica, and has been financially supported by the National Academy of Sciences of Ukraine and COLCIENCIAS through Project n° RC 200/93, and by the INTAS programme 1993-2779.

6 REFERENCES

1 J. R. van Asperen de Boer. “Infrared reflectography of Paintings using IR vidicon TV Systems,” *Studies in Conservation*, N° 14, pp. 96 - 118, 1969

2. S. Fletcher. “A Preliminary Study of the Use of IR Reflectography in the Examination of Works of Art on Paper,” *ICOM 7th Triennial Meeting*, (Copenhagen), pp. 62-64, 1984
3. Proc. of the “4th International Conference Non-Destructive Testing of Works of Art”, Berlin, 1984.
4. V. B. Markov, G. I. Mironjuk, I. G. Yavtushenko, *Holography and its Application for Museum work*, UNESCO Press (Paris), 1984
5. Yu. Denisyuk, “Art holography in thick materials based on Lippmann photographic plates” *J. of Techn. Phys.* Vol. 48, pp.1683-1687, 1978
6. M. Lehmann. “Three-Dimensional Display,” in: *Handbook of Optical Holography*, ed. by H. J. Caulfield, Academic Press, New-York, pp. 447-462, 1979
7. *Holographic Non-destructive Testing*, ed. by Robert K.Erf, Academic Press, New York - London, 1974.
8. N. G. Vlasov, V. M. Ginsburg, et all, “Interferometric Holographic Methods for the Definition of Optimal Temperature-Humidity Regimes for Museum Items Storage”, *Sov. Phys. Dokl.*, Vol. 225, pp. 1312-1315, 1975
9. S. Amadesi, F. Gori, R. Gralla, G. Guattari. “Holographic Methods for Painting Diagnostic,” *Applied Optics*, Vol.13, pp.2009-2013, 1974
10. S. Amadesi, D’Altorio, D. Paoletti. “Sandwich Holography for Painting Diagnostic,” *Applied Optics*, Vol.21, pp. 1889-1890, 1982
11. D. Bertani, M. Cetica, G. Molesini. “Double Exposure Holography on the Giberti Panel “THE LIFE OF JOSEPH,” *Proc. SPIE*, Vol. 370, pp. 255-258, 1982
12. J. Defang, D. Paijun, et all. “Holographic Non-destructive Testing for Cultural Relics,” *Proc. SPIE*. Vol. 370, pp. 259-265, 1982
13. V. B. Markov, A. S. Monchak, “Interference-holographic method of the museum item conditions analysis,” Proc. of the Conference *Applied Holography*, 1984, L’vov, p.17
14. G. Accardo, P. de Santis, F. Gori, et all. “The Use of Speckle Interferometry in the Study of Large Works of Art,” *Journal of Photographic Science*, Vol.33, pp. 174-176, 1985
15. G. Gülker, K. Hinsch, C. Helscher, “Electronic speckle pattern system for in situ deformation monitoring on buildings”, Optical Engineering, vol. 29, No 7, 816-820, 1990
16. P. Boone and D. Van Nieuwenburgh, “Optical analysis of environmental influences on artefacts” in *"Optics for protection of man and environment against natural and technological disasters,"* pp.183-187, Elsevier-Amsterdam, 1993
17. J. Paoletti et all. “Automated DSPI contouring in artwork surface inspection”, *Optical Engineering*, Vol. 32, pp. 1348-1353, 1993
18. R. Jones, C. Wykes, *Holographic and Speckle Interferometry*, Cambridge U.Press, Cambridge, 1983.
19. John F. Ready, “Effects of High-power Laser Radiation”, Academic Press, New-York - London, 1971, Ch.3.

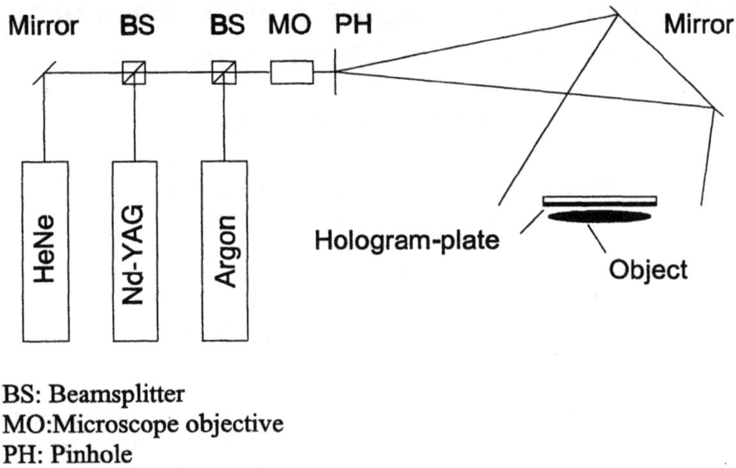

Fig. 1 Denisyuk configuration for single beam color-holography. The three wavelengths (488nm, 532nm, 633nm) were coupled into a spatial filter. A spherical mirror controls beam incidence and divergence. Exposures were made simultaneously with three wavelengths.

To minimize exposure time, the hologram was recorded simultaneously with three wavelengths (b:488 nm, g:532 nm, r:633 nm) [16]. For the blue line an argon (Spectra Physics 2000), for the green a Nd:YAG (Adlas DPY 452 II) and for the red line a He-Ne laser (NEC) was used. The intensities of the three lasers were adjusted to achieve an ideal white in the holographic image for later reconstruction with a halogen lamp.

3 Color measurement

To measure the color rendering of the holographic image, a Kodak color test target was used. The recording and reconstruction of the hologram is described above. The observation angle was normal to the hologram plane and the distance 20 cm. In reconstructing the hologram, the diffracted light was collected with a telescope and coupled into a glass fiber bundle. To analyze the spectral distribution of the diffracted light a grating-spectrometer (Amko) was used. With colorimetric software the color coordinates were calculated for each measured spectra.

Fig. 2 demonstrates the measured color differences ΔE (CIELAB) [17] including the difference of lightness. The color rendering of the photo shows a small visible difference to the original color test target. The result of the measurement (photo) is in good agreement with other authors [18].

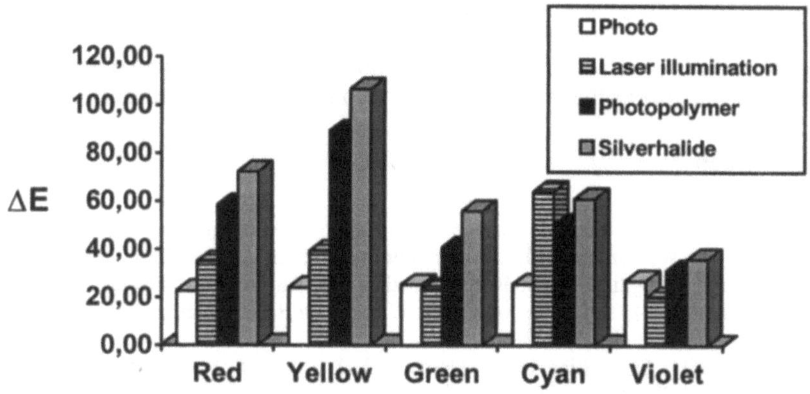

Fig. 2 Color differences ΔE (CIELAB) for the reproduction of five colors: a photo (white light illumination), the original target (laser illumination) and two different holograms on silverhalide and photopolymer material (white light replay).

With laser illumination (three wavelengths) it is not possible to achieve a color rendering comparable to white-light illumination for all colors. The color difference depends on the spectral distribution of the illuminated test color and on the used laser wavelengths.

The holographically replayed colors show the largest color differences. Significant is the decreasing of saturation in the holographic image. The colors reproduced by the hologram on photopolymer are more saturated than the colors on silverhalide. This is caused by the better SNR of photopolymer and by the smaller peak-bandwidth. The thicker emulsion of photopolymer ($d=20\mu m$) results in a smaller peak-bandwidth for white-light replay. The use of more than three wavelengths for color holography would improve the color rendering with respect to the chroma, but lightness and saturation would decrease because of worse SNR due to color multiplexing. A holographic image will always have less brightness than the original object. The hologram works like an optical filter and most of the intensity is lost.

4 Conclusion

We have studied the properties of white-light illuminated color reflection holograms with respect to color rendering. It was demonstrated that silverhalide (PFG) and photopolymer (HRF) are suitable for producing high quality color holograms. Very high diffraction efficiencies and low noise in the holographic image of a true-color hologram are possible. The color rendering of the laser illuminated color test target shows that the proper choice of wavelength selection depends on the spectral reflectance of the objects. To improve the color rendering with holography, more than three wavelengths are necessary. But at the moment there is no material which is able to record such a large number of different gratings in one layer without great losses in diffraction efficiency and increasing noise. To achieve more saturated colors it would be an improvement to make a reflection copy (photopolymer) of a transmission hologram (photopolymer). This will increase the diffraction efficiency. A better SNR will be achieveable because the intensity of reference- and object wave can be optimally adjusted. One problem is the shrinkage of the photopolymer. The heating after exposure attacks the layer and a laser reconstruction is not uniform.

References

1. G. von Bally, F. Dreesen, A. Roshop, E. de Haller, G. Wernicke, N. Demoli, U. Dahms, H. Gruber, and W. Sommerfeld, „Holographic methods in cultural heritage preservation and evaluation", in Optical Methods in Biomedical and Environmental Sciences (H. Ohzu and S. Komatsu, eds.), 3 of Series on Optics Within Life Sciences, Elsevier Science Publishers B.V., Amsterdam (1994) 297.
2. G. von Bally, D. Vukicevic, N. Demoli, H. Bjelkhagen, G. Wernicke, U. Dahms, H. Gruber, and W. Sommerfeld, „Holography and Holographic Pattern Recognition for Preservation and Evaluation of Cultural-Historic Sources", Naturwissenschaften 81 (1994) 563.
3. P. Greguss,"The Historical Way to Holoendoscopy in Medicine and Technics", Laser and Optoelektronic 27(6) (1995) 57.
4. H. I. Bjelkhagen, J. Chang, K. Moneke, „High-resolution contact Denisyuk holography", Appl. Opt. 31, 1041-1047, 1992.
5. R. Smith, „Holographic Center High Mounted Stoplight", Proc. SPIE, vol. 1461, pp. 186-198, 1991.
6. A. Ramsbottom, S. Sergeant and D. Sheel, „Holography for Automotive Head-Up-Displays", Proc. SPIE 1667, 1992.
7. R. L. Fisher, „Design methods for a holographic head-up display curved combiner", Opt. Eng., vol. 28 no. 6, pp. 616-621, 1989.
8. J. R. Magariños and D. J. Coleman, „Holographic mirrors", Opt. Eng., vol. 24, no. 5, 1985.
9. D. J. McCartney and D. B. Payne, „Position-tuneable holographic filters in dichromated gelatin for use in single-mode-fiber demultiplexers", Optics Letters, vol. 10, pg. 303, 1985.
10. J. L. Salter and M. F. Loeffler, „Comparison of dichromated gelatin and DuPont HRF-700 photopolymer as media for holographic notch filters", Proc. SPIE, vol. 1555, pp. 268-278, 1991.
11. X. Ning and J. Masso, „Multi-line holographic notch filters", Proc. SPIE, vol. 1545, 125-129, 1991.
12. R. W. Brandstetter, N. J. Fonneland, „Photopolymer Elements for an Optical Correlator System", Proc. SPIE, vol. 1559, pp. 308-320, 1991.
13. N. J. Phillips and C. A. Barnett, „Micro-optics studies using photopolymers", Proc. SPIE, vol. 1544, pp. 10-21, Sept., 1991.
14. B. Omar, F. Clube, M. Hamidi, D. Struchen and S. Gray, „Advances in Holographic Lithography", Solid State Technology, pp. 89-93, Sept., 1991.
15. A. A. Ward, J. C. W. Newell and L. Solymar, „Image blurring in display holograms and in holographic optical elements", SPIE Proceedings 600, 57-65.
16. G. von Bally, F. Dreesen, V. Markov, A. Roshop, E. de Haller, „Registration of color holograms on PFG-03-C monolayers", Journal of Technical Physics (russian) 21, 76-80, 1995.
17. CIE, International Commission on Illumination, Colorimetry, 2nd ed., Pub. 15.2, Vienna, 1986.
18. Karlheinz Keller, „Science and Technology of Photography", VCH Verlagsgesellschaft, Weinheim, 1993.

Three-Dimensional Microscopy with a Numerical Optical Low-Coherence Holographic Technique

E. Cuche[1], P. Poscio[2], Ch. Depeursinge[1].

[1] Institut d'Optique Appliquée, EPFL, CH-1015 Lausanne, Switzerland
[2] Institut de Recherches en Ophtalmologie (IRO), CH-1950 Sion, Switzerland
E-mail: etienne.cuche@ioa.dmt.epfl.ch

Abstract. We present here a numerical application of Optical Low Coherence Holographic Microscopy (OLCHM). An object is illuminated by a low coherent source and the backscattered light interferes with a reference wave at the exit of a Mach-Zender interferometer. The resulting hologram is recorded by a CCD camera and numerically reconstructed. The precise control of the reference's optical path length allows the selection in depth of the backscattered light with a depth resolution of about twice the coherence length of the source.

1 Introduction

Observation of microstructures in three dimensions is of great interest in numerous domains of the life sciences and humanities, archaeology in particular. Because of their low cost and their easy use, optical techniques are appropriate for this purpose, especially when non-destructive techniques are necessary. However, except for conventional holography whose practical application can be complicated, very few optical techniques actually enable a real 3D imaging. A promising way consists in performing the selection of the so-called "first arriving light" by means of low coherence holographic methods [1-4]. We present here an application of such a technique combined with recent developments of numerical holography [5].

2 Method

The experimental setup is shown in fig. 1a. After enlargement of the spot size by a beam expander (B.E.), a polarizing beam splitter (PBS) separates the light. On one hand, the vertical component of the polarized beam is reflected in the direction of the object and a magnification optics (a single lens of 30 mm focal length) collects the backscattered light. On the other hand, the horizontal component crosses the PBS and forms the reference wave which interferes with the object wave at the exit of the interferometer. The resultant hologram is recorded by a standard CCD camera with $Nx \times Ny = 512 \times 768$ pixels of $\Delta x \times \Delta y = 6.8 \times 6.8\ \mu m^2$ area. For low coherence investigations, a pulsed Ti:sapphire laser has been used and a delay device has been introduced in order to control the reference's optical path length. It consists of a single prism whose position can be accurately adjusted by a micrometer screw. Results in fig. 2 illustrate the importance of this delay device.
Finally, the image is numerically reconstructed by a personal computer. The used algorithm consists of a discrete formulation of the Fresnel Transform [6]:

Finally, the image is numerically reconstructed by a personal computer. The used algorithm consists of a discrete formulation of the Fresnel Transform [6]:

$$\Psi(m\Delta\xi, n\Delta\eta) = \exp\left\{\frac{i\pi}{\lambda d}(m^2\Delta\xi^2 + n^2\Delta\eta^2)\right\} DFT\left[I_H(k,l)\exp\left\{\frac{i\pi}{\lambda d}(k^2\Delta x^2 + l^2\Delta y^2)\right\}\right]_{m,n} \quad (1)$$

Where $k,m=1..Nx$; $l,n=1..Ny$ are integers, $\Delta\xi$ $\Delta\eta$ are the sampling intervals after the discrete Fresnel Transform (see refs. [5,6]), λ and d are the wavelength and the reconstruction distances respectively. *DFT* is the Discrete Fourier Transform operator which can be very quickly computed using FFT algorithms. $I_H(k,l)$ is a matrix of $Nx \times Ny$ elements which results from the sampling of the hologram intensity by the CCD camera.

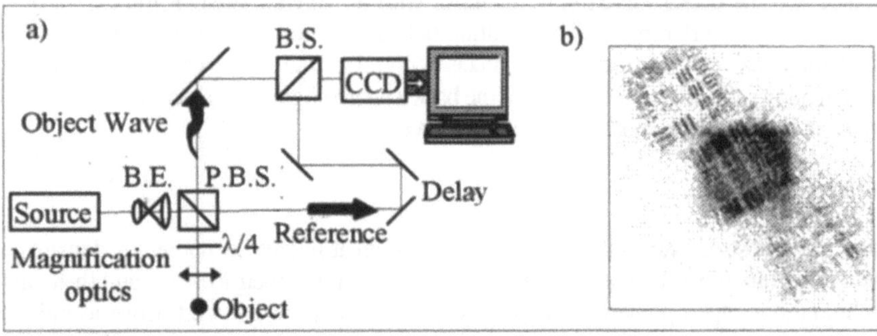

Fig 1. a) Schematic of the numerical OLCHM imaging setup. b) Typical result of numerical reconstruction

Fig. 1b presents a typical result of numerical reconstruction using equ. 1 with a reconstruction distance d=55 cm. The hologram was recorded with a HeNe laser and the groups 2 and 3 of the USAF test target as objects. No magnification optics were used here. The computation time was less than 4 seconds for 512 × 512 points with a 200 MHz Pentium Pro processor.

Three different terms are visible in fig.1b. The zero order diffraction appears in the centre of the image. In the top left corner, one can observe the real image and, symmetrically with respect to the centre, a blurred twin image. These three terms are spatially separated because an angle was introduced between the reference and object waves at their incidence on the CCD camera.

3 Results and Discussion

Images in fig. 2 have been reconstructed from holograms recorded with the Ti:sapphire laser and the character "4" of the title of group 4 of the USAF target as the object. A single lens of 30 mm focal has been used as magnification optics. Firstly, the delay device was set on the optimal position for the considered object. Then, six holograms have been recorded after displacements, with 10 μm steps of the delay device, backward and forward with respect to its initial position. We can see in the corresponding numerically reconstructed images that the character "4" disappears along both directions for displacements higher than 20 μm.

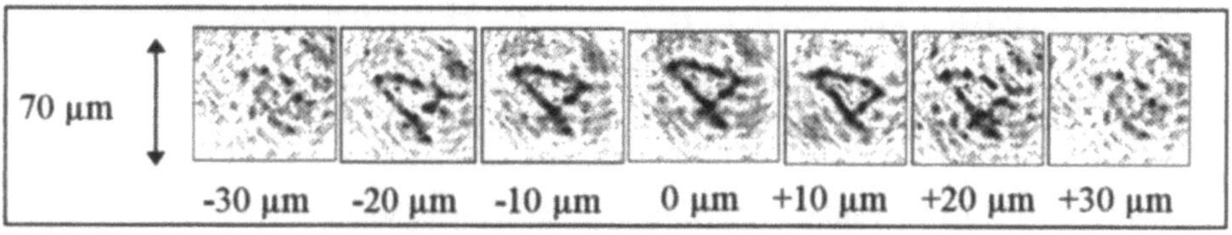

Fig. 2. Depth resolution with a Ti:sapphire laser: Reconstructed images for seven different positions of the delay device.

From this result, the depth resolution of our technique can be estimated to be about 40 µm. This means that the part of the backscattered light which has been recorded in the hologram comes from a volume in the object area whose thickness is equal to less than twice the coherence length of the source. Indeed, the 80 fs used pulses are 24 µm long.

4 Conclusion

The numerical reconstruction algorithm has been successfully tested. The results obtained with a pulsed Ti:sapphire laser have shown the technique to be efficient for 3D microscopy with a depth resolution equal to twice the coherence length. Further development of a specialized hardware and the use of superluminescante diode should enable a video frequency reconstruction with a low cost technique. This opens a wide range of interesting perspectives for new applications of the numerical OLCHM, in biomedicine as well as in numerous other domains where the visualization of 3D scenes at the microscopic scale is necessary.

5 Acknowledgments

This work was supported by the Swiss National Foundation (grant FN 2100-042323.94/1) and the UNIL-EPFL grant (New optical methods in ophtalmology). The authors thank Frédéric Bevilacqua and Olivier Coquoz for interesting remarks about the manuscript.

References

[1] E. Leith, C. Chen, H. Chen, Y. Chen, D. Dilworth, J. Lopez, J. Rudd, P.-C. Sun, J. Valdmanis, G. Vossler, J. Opt. Soc. Am. A 9 (1992), 1148.
[2] S. C. W. Hyde, N. P. Barry, R. Jones, J. C. Dainty, P.M.W. French, M. B. Klein, B. A. Wechsler, Opt. Lett. 20 (1995), 1331.
[3] O. Coquoz, R. Conde, F. Taleblou, Ch. Depeursinge, Appl. Opt. 34 (1995), 7186.
[4] W. S. Haddad, D. Cullen, J. C. Solem, J. Longworth, A. McPherson, K. Boyer, C. K. Rhodes, Appl. Opt. 31 (1992), 4773.
[5] U. Schnars, W. Jüptner, Applied Optics 33 (1994), 179.
[6] L. P. Yaroslavskii, N.S. Merzlyakov, Methods of Digital Holography (Consultant Bureau, New York, 1980).

Holographic Investigation of the Human Jaw Bone Under Functional Loads

L. V. Tanin, I. L. Drobot, A. S. Artushcevich,
A. S. Rubanov, A. S. Naumovich

Institute of Physics, Academy of Sciences of Belarus
70 F. Skaryna Prospect, Minsk 220072 Belorussia
Tel. +375-172-394419, Fax: +375-172-393131,
E-mail: hololab@bas33.basnet.minsk.by

Abstract. The article deals with the determination of deformations of the human jaw bone under functional loads using holographic interferometry. Quantitative evaluation of interferograms of the mandible has been carried out. The deformations in horizontal and frontal planes on the level of the alveolar process, body, and lower edge of the jaw were determined.

1 Introduction

Holographic interferometry is one of the most informative methods for determining the deformation of physical bodies. This method is used in stomatology for studying the deformation of sets of teeth substitutes, of plastic models of the mandible, and of bones of the facial skeleton [1, 2, 4]. In these earlier investigations, loading experiments were performed without taking into account the mechanical forces resulting from the contraction of muscles which move the lower jawbone. The purpose of our work was to study the spatial deformations of the lower jawbone, arising from biting and swallowing motions. The research results are of significance for the development and planning of methods of treatment in surgical and orthopedic stomatology.

2 Experiments

A special loading device was developed, which enables precise simulation of the traction of all groups of chewing muscles. Thus, direction and absolute force of each group of muscles were taken into account.

In the experiment, we used a holographic setup with counterpropagating beams with a recording plate fixed directly to the object. The light source was a He-Ne laser, generating single-mode radiation with a wavelength of 632,8 nm. Recording of interferograms was made on photo-plate type PFG-03.

3 Evaluation

The qualitative evaluation of deformations of the lower jawbone was made according to the spatial distribution of the interferometric fringes. Main criteria were the spatial frequency of the fringes and their direction. The fringe density was dependent on the degree of object deformation. It is possible to make a judgement about the distribution of the deformation based on the direction of the fringes since interferometric fringes always pass perpendicular to the direction of the main deformation [3].

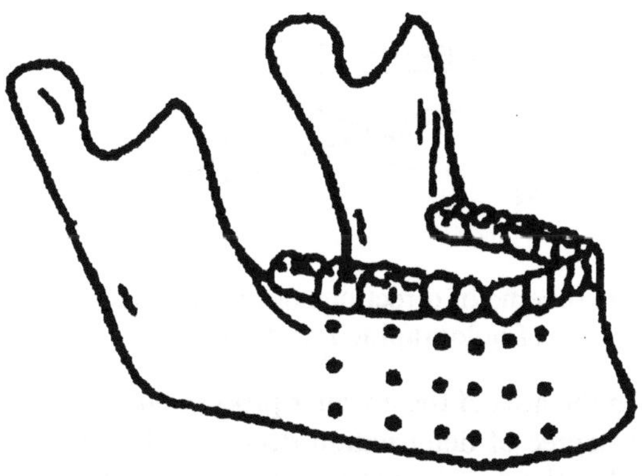

Fig. 1. The points on the surface of the mandible in which the functional deformation was investigated

A qualitative evaluation of the holograms of the lower jawbone demonstrated that the object tested showed various deformations in horizontal as well as in frontal planes. On the basis of the distribution of the interferometric fringes it is possible to conclude that the more significant deformations are observed in the area of the teeth and the alveolar process, compared to the area of the bottom edge of the jawbone. In the chin region, the deformation is less than in the area of the corpus and at the angle of the jaw.

For a more exact estimation of the deformations, a quantitative evaluation of the interferograms was performed. We used the technique of absolute orders of fringes. As the object can be presented as a flat body in a first approximation, an interferometric setup with symmetric directions of observation was chosen. We counted the fringes starting from a zero point, which was chosen at the place where the hologram plate was fixed. After evaluation of the interferogram at each of the defined points on the surface of the lower jawbone from 4 directions of observation, we calculated the magnitude of deformations of the object for 3 coordinates according to the sensitivity equations (see fig. 1 and [3]).

4 Results

As a result of the experiments we have obtained deformation fields of the mandible on coordinates X1 and X2 (fig. 2). Maximum deformation is observed along the axis X1. It turned out that the more significant shift corresponds to the basis of the jawbone in the field of the second molar and gains 3,66 μm.

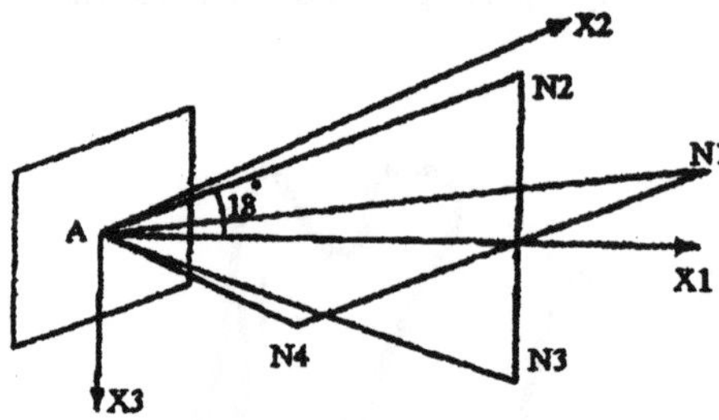

Fig. 2. The scheme of directions of lighting and observation of holographic interferograms

The deformation in the area of the alveolar process on the same level was smaller (2,89 μm). Thus, the degree of deformation decreased towards the direction of the chin area. It correlated to the distribution of loads in the area of the alveolar process, the corpus, and the basis of the jawbone. The deformation on the axis X2 was less prominent. Its maximum corresponded to the projections of the second molar (0,41 μm) in the area of the corpus of the mandible at the level of the canine (0,77 μm). In the area of the basis, the deformation had a negative sign (-0,19 μm) according to the projection of the second premolar.

5 Conclusions

Quantitative processing of holographic interferograms of the mandible, obtained at the contraction of chew muscles of both groups, has allowed us to investigate how the deformations in the area of alveolar process, basis, angle and symphysis are distributed in the horizontal and frontal planes. The maximum deformations were observed in the area of the alveolar process and the jawbone angle in both planes. The deformation decreases significantly in the frontal area of the jawbone.

References

1. M.A. Azizov, V.G. Bachtin, S.P. Poluchin, Stomatology **6**, 66-68 (1985).
2. V.S. Siniakov, M.I. Kogevnikova, Successes of Physiological Sciences **19,** 109-116 (1988)
3. Ch. Vest, Holographic interferometry. 540 (1982).
4. J.C. Ferre, R. Ligoux, J.L. Helary, J.Anat. Din **7,** 193-201(1985).

Other Analytical Techniques (Material Analysis, Dating, Localization, Biomolecular Methods, etc.)

3D Computerized Tomography: Synergism Between Technique and Art

B. Illerhaus, J. Goebbels, H. Riesemeier

Bundesanstalt für Materialforschung und -prüfung (BAM), Unter den Eichen 87, D-12205 Berlin, Germany

1 Introduction

Since 1970, the well known X-ray technique has developed into the new branch of computerized tomography (CT) [1]. First developed for medical purposes, this method is used nowadays also in the field of cultural objects and objects of art. The interesting results give rise to new interpretations from the archaeologist's point of view. In contrast to normal X-ray imaging, computerized tomography allows the calculation of density in each point of the investigated object. Measurements started with two dimensional cross slices (2D-CT), but today a three dimensional object can be investigated in its entirety at the same time (3D-CT). Whereas technical objects consist mainly of one or two known different but similar materials, the range of densities within archaeological objects is much more widely spread and often unknown. The investigation of such interesting cultural objects raises the research efforts usable in the technical branch.

2 X-ray investigations inside museums

One hundred years after the discovery of X-rays, most restoration departments are equipped with their own X-ray tubes. Pictures are investigated to show their historical quality as well as to reveal their painting process and to ensure the genuineness of objects of art. The advantage of X-rays is the penetration of the objects with nearly no harm; the only exception is the destruction of any thermoluminescent information. But X-ray images are always only a two dimensional projection of a three dimensional object: the depth information is missing or a dense object may hide others behind it. Furthermore, parts of the object may require very different film exposure times and thus may not be recorded at one time. CT investigations, on the other hand, could provide a fully three dimensional (virtual) replica of the object, showing even invisible interior parts in detail.

At the moment, none of the corresponding museums is equipped with a CT-system because these have been developed for medical or precise technical applications and the costs are quite high. Therefore, the investigations presented here were done on medical scanners (restricted to light materials) or on scanners developed for nondestructive material investigations of technical applications, such as ceramic or metallic parts. But it would be very useful to have much more precise information on archaeological findings before starting any restoration, because any changes to these unique pieces are irreversible.

3 The principle of computerized tomography

A collimated (pencil) X-ray beam penetrates the object of investigation. As in the normal radiographic case, it will be attenuated proportional to the length and to the density of the absorbing material. (correctly: proportional to the atomic number.) After an additional collimation (or filtering by light materials) to suppress scattered radiation, the intensity is measured, digitized and stored in the computer together with the coordinates. As required by the planned spatial resolution, the measurement is repeated step by step, then turned for a small angle and then measured again (fig. 1). After a 180° turn the measurement is complete, because the rest is symmetric. The energy used for the radiation should be high enough to penetrate the object in every direction to a measurable part, otherwise image artifacts would result. The collected data are normalized, inverted and taken to logarithm in the computer to obtain the underlying measuring value: the linear X-ray absorption coefficient. From this data set, the density in each object point is calculated by filtered backprojection [2].

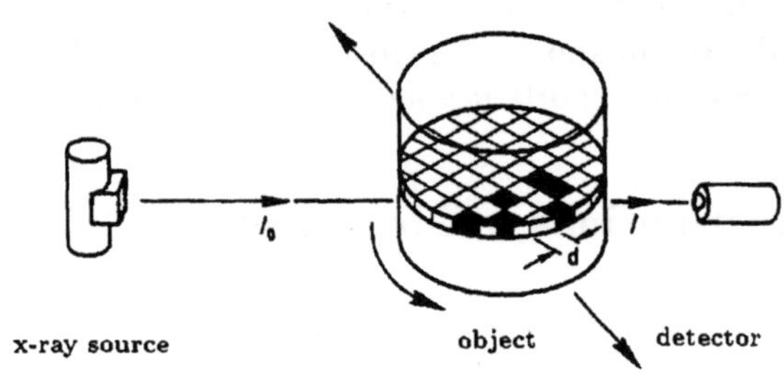

Fig. 1. The principle of 2D-CT

4 The tomographic image

The result of this calculation is the tomographic image which gives a cross sectional view. Each element (pixel) contains the calculated density of a volume element (voxel) mainly defined by the collimator opening in cross sectional plane and perpendicular to it. Usually the density value is represented in the image by grey levels in 256 steps (up to 4096 possible). Black is equivalent to air, white is set to the highest occuring density. If the object contains very dense materials (e.g. silver) as well as light materials (e.g. wood), the grey levels have to be written in more than one byte or the different materials have to be stored in different images. The human eye is not able to distinguish between more than twenty grey levels. For showing small density variation in an image there are two possibilities: the interesting density region is enhanced to cover the whole 256 values, or a colour table is used. Using a colour conversion table must result in an image showing the same dark to light behaviour as the grey level representation to ensure an easy correlation between colour and density. A changing step between two colours should only occur when there is a definitely different material.

5 Measurement quality

The quality of a CT measurement is defined by the density and the spatial resolution. The first is proportional to the square root of per voxel measured X-rays. Thereby the energy and intensity of a suitable X-ray tube is preset: the energy for maximal penetration length in steel is: 200 kV = 2 cm; 400 kV = 10 cm; 1 MeV = 15 cm; 12 MeV = 40 cm. The spatial resolution is mainly dependent on the geometrical set-up. It will be measured at a sharp edge in the reconstructed image. A rough geometrical estimation can be deduced from the opening of the used collimators, or from the size of the detectors and the focal spot size. The source should have a well defined spot with more then 90% of the output radiation inside.

6 Technical set-ups

In practice, the source-one-detector set-up has changed to several possible multi-detector arrangements. The measuring time is shortened according to the number of used detectors. A medical tomograph needs only a few seconds per slice measurement. In the technical field, developments have lead to a two dimensional detector screen, equivalent to up to 1000 by 1000 detectors and a fully three dimensional investigation.

7 Three dimensional tomography (3D-CT)

If neither the inner structure nor the location of the estimated error is known, an object has to be tested in total. This can be done by adding adjacent two dimensional CT slices until the total object is covered. This is done in the case of medical scanners. The other possibility is the 3D-CT: the object is X-rayed in total as a normal radiographic image on a two dimensional screen (fig. 2). After a full turn, a volumetric image can be calculated (e.g. by the Feldkamp algorithms [2]). Here, a collimation in front of the detector is no longer possible. Therefore the scattered radiation will influence the image and reduce the density resolution. Furthermore, the active volume of a detection cell within a two dimensional screen is smaller than in the case of one dimensional lined arrays. The great advantage of the 3D-CT is that after one turn, the total object can the copper solder is uncovered. In former times, only the front panel of these iron objects was restored, and the back was crudely filed down. Any information about the manufacturing was lost. With CT, the used soldering technique can be studied without any destructive damage. On the computer screen, the silver and copper parts can be shown separately. In the side view (fig. 4), the very thin copper layer between the upper and the side part, and between the side and the bottom part is clearly visible. At the rounded end, the copper surplus is covering the total inner surface of the iron side wall.

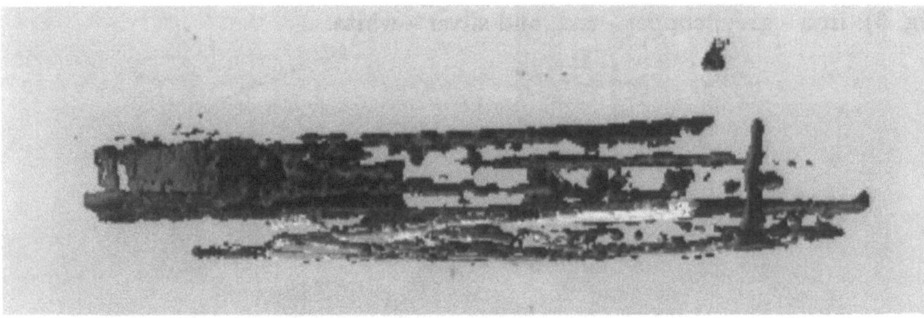

Fig. 4. Copper and silver parts of fig. 3 in side view

Furthermore, one of the original two brass rivets fixing the leather belt is shown. The small copper coloured component embedded in the rust is its shifted back-side counter part. Showing all metallic parts and looking from the straight end, the cavity where the belt was plugged in can be inspected. A tomographed finding can be virtually restored and carefully analysed prior to any restoration. The knowledge about the (inner) setup of all parts will preserve the original and will allow the restorator to get as close as possible to the original appearance. Sometimes an object may be revealed as very fragile or having large cracks so that a real restoration may be impossible. The analysis of the virtual images could then be an alternative. On the other hand, all results from CT are reproducible, screenable and non-destructive.

8.2 Brass and copper objects

Brass statues hide hints to their construction inside. Exact studies of the joining points and of the wall thickness reveal the quality of the technique used [5]. Due to the scanning of object geometry with CT, a comparison of different statues and their documentation is possible. Most outer surfaces have been treated after casting, but the inner surface stayed the same. The production method can be deduced from several remains such as the sand kernel, the joinings of kernel parts, or the kernel supports.

Herm, representation of an inner, non-accessible surface

After a casted statue has been tomographed, both surfaces, the outer and the inner, are reconstructed and can be shown by means of image processing. A herm found in a shipwreck near Mahdia, Tunesia [8], which was restored in the Rheinisches Landesmuseum Bonn, was investigated with 155 single tomographic slices in 1-2 mm intervals by BAM's large scale scanner. The statue is made in the head region from only one kind of bronze, and there are nearly no remaining parts of the sand kernel. Furthermore, this head has several holes (e.g. the eyes) to get a look inside. Therefore, there are no unknown parts expected to be inside. The reason for the total inspection of the head region was the chance to get a portrayal, seen from the outside, of the total inner surface, allowing a direct comparison to the outer surface. A new software tool had to be developed. The inner surface was calculated by omitting the outer and inverting the normal vector of the inner surface (fig. 5): The sand kernel originally had nearly the same shape as the outer surface, especially in the region of the headscarf, resulting in a thin walled casting. Nevertheless, some characteristic outer parts are not preshaped within the kernel. This means that these parts were added to the prefabricated kernel as additional wax. Surprisingly, there was a crack in the kernel arround the whole left part of the face in the height of the mouth, tilting the upper head kernel slightly.

There is a second herm with nearly the same appearance: the Getty herm (79.AB. 138, Collection of the J. Paul Getty Museum, Malibu, California) in the J. P. Getty Museum. There are some hints that both hermes were produced from standard forms [8]. Therefore it is desirable (and planned) to scan the Getty herm in the same way [18]. The produced geometries will then allow an exact comparison; especially the inner surfaces will most likely reveal any similarities. Will it be possible to deduce that both were cast by the same manufacturer?

Restoration problems, a copper statue from Ashur

A small statue from Tell Agule (Iraq) (fig. 6 shows the outer surface and the inner sand kernel of the head) was found strongly corroded. Besides the scientific interest in the casting technique, the underlying original surface had to be found. In an aggressive surrounding, copper tends to corrode in total, generating a collective expanding surface with great cracks between corroded and uncorroded material. Here, every single slice of the measured head region had to be inspected manually. Arbitrary slices were used additionally to clarify whether the outer surface hides any information, e.g. is there really no mouth? On the other hand, careful inspection showed that the bulge representing the eyebrows is only visible in the outer surface; a blast off of the corrosion layer would destroy this information. Although restoration nowadays is as careful as possible, CT investigations may help to find the right decisions.

Mixed casting, the animal bearer VA5010

Two brass statues were found in a treasure in Ashur and are nowadays part of

Fig. 5. The Mahdia Herm, outer and inner surface

the ongoing exhibition in the Vorderasiatisches Museum Berlin [6, 7]. Although found side by side and showing great artistic knowledge in most parts, the face of the second turned out badly. Because the colour of the front part is different, an analysis was made. This part of the face is nearly pure copper. There are two explanations. First, this may be a product of corrosion. Salt-water can leach the tin from a brass casting. Heavy corrosion could have produced the cracks visible from outside (compare to the statue above). The second possibility is an additional casting with copper to the cold (but partially failed casting?) statue.

Although the head was inspected with a 3D-CT equipped with an 200 kV micro focus tube, the penetration limit was nearly reached. Copper and brass are normally easily separable by their density in a tomographic image. But in this case, the beam hardening effect predominates, especially because the most outer material is the less dense material. The beam hardening effect leads to an overestimation of the density; in this case it equates some outer parts of the copper with inner brass parts. This problem is easily solvable for parts consisting of one material, but is not yet solved for two or more materials. This problem gave rise to a project sponsored by the EC to develop an iterative correction software. In the present case, this problem has been solved by separation by hand. Afterwards, the underlying brass surface was generated. Fig. 7 shows both surfaces in comparison. In the brass, nose, mouth and chin are already formed, and the eyes are comparable to the second statue. Thus it seems that the first brass casting had not been a failure. The CT provides the fact, but the explanation and interpretation still have to be done.

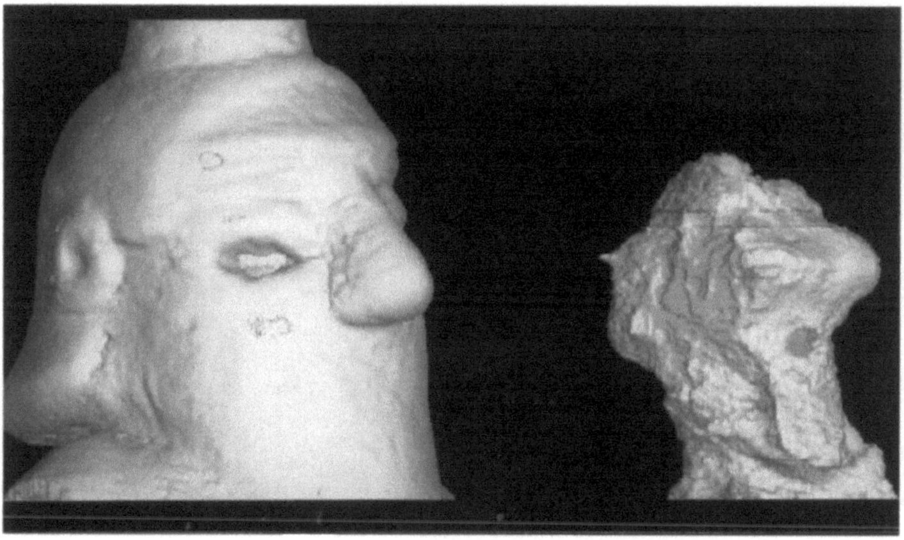

Fig. 6. A small statue from Tell Agule, outer and inner surface

9 Teje head

Some historical art objects have been changed in the course of time; the original appearance seems lost and today's onlooker can only guess what the original looked like. In the simplest case, the original is hidden underneath. If the current outer surface is as historical as the underlying one, it is not possible to restorate the latter. This is true in the case of the portrayal head of Queen Teje (Collection of the Ägyptisches Museum Berlin)[9]. Some damaged spots in the covering hair cap showed an underlying silver cap and gold jewellery. Although X-ray images taken long ago had revealed very complex structures of gold and silver, it was not possible to draw a map of the original sight.

The head has a diameter of about 8 cm. Thus the resolving of details requires a high spatial resolution for CT investigations. On the other hand, the gold parts are very dense; too dense to be penetrated in each direction by a 420 kV X-ray tube, thus producing artifacts. Furthermore, the spatial resolution of the used CT system is limited to 0.5 mm (the micro-CT with 380 kV is still a project). Thus an exact CT reconstruction is not yet possible. Nevertheless, we measured 93 single slices, combined them to a 3D set, giving a rough estimation of the inner set-up. For a more precise image, a step width of 0.1 mm in each (of three) dimensions would be necessary. With the available machine this would last 130 days. Single tomograms, consisting of slices in the range of forehead and eyes showed: the 256 linear grey levels are not enough to differentiate all materials with a good density resolution. Here, only the light weight materials are shown and therefore the thickness of the silver cap (blue) is overestimated [20].

The basic form of the head was made of two parts of different wood; the

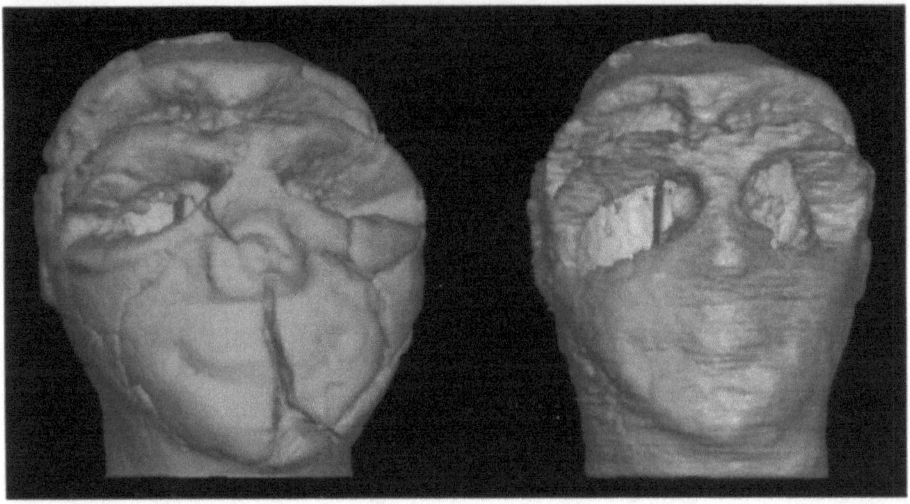

Fig. 7. Animal bearer VA5010, left as seen, right copper face removed

different structure of the veining is visible. Between both wood parts, at the border of the rear part, a zone of material with higher density (orange) in the wood exists, probably resulting from an organic glue. All around the silver cap (blue) and on both sides of the face, the hair cap consisting of glue and tissue material is closely fitting. The density of the tissue is very low, but due to the addition of glue, the density becomes comparable to wood. A close look reveals the ears of Queen Teje as a triangle in front of the silver tin, showing the same veining. In the middle of the image, the thorn originally bearing the sign of divine dignity is cross sectioned. (This wood panel was stored separately in the museum but is now again added to the head.) The second tomogram gives a cross section through the gold jewellery (white). The produced artifacts disturb the image. At the left side, one of the golden fixing nails of the ear jewellery is visible. In a vertical slice through all CT slices, the veining parallel to the face surface is shown.

The surfaces of different materials were calculated from the existing data: At first the silver and the gold parts, then the wood part of the face leaving out the glued tissue parts. Subtracting the front part of the face from the total data set yields the hair part. The glue-containing surface between both parts and the tenon can thus be inspected. Any sight of the surfaces can now be constructed on the computer screen. Parts may be shown (semi-) transparent to clarify the spatial, geometrical set-up. One view (fig.8) shows only the gold jewellery and the wooden face to show the hidden surface of the wood part and the fixings of gold. The next image gives a possible view of the total head as it might have looked during the reign of Queen Teje. The wood is coloured more red as is normal for recently-cut yew wood.

Striking are the two golden "Uräus" snakes following the edge of the silver cap. Because the outer form of the silver cap resembles the traditionally styled Egyptian hair, the face appears slimmer and longer. With our method, the question whether the cap was artificially oxidized from the beginning cannot be clarified. The shimmering blue-black would be near to the natural hair colour, corresponding to the yew wood used. Or should the splendid shining polished silver be prefered? This reconstruction from CT data will lead to new interpretations in the historical view of Queen Teje.

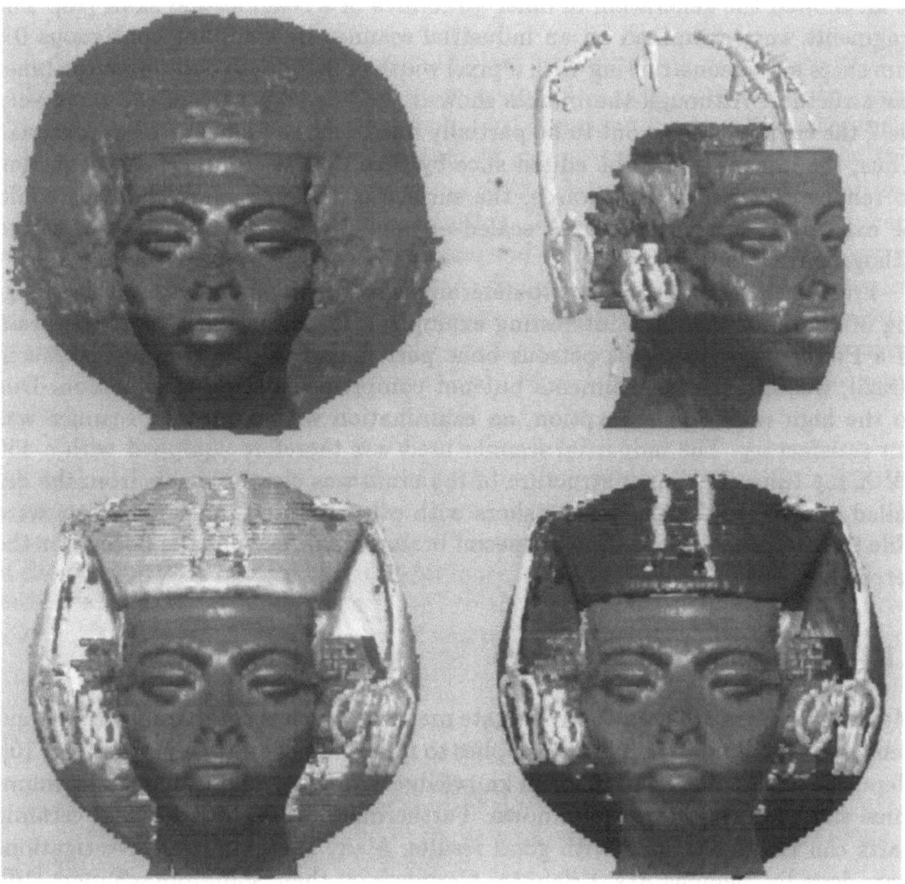

Fig. 8. Head of a portrayal head of Queen Teje. top left: current outer surface, top right: only the gold parts of the hair cap shown, bottom left: possible restoration with shiny silver cap, bottom right: with dark blue oxydation

The data set yielded by tomography was used to generate a double of the uncovered state by stereo lithography (LBBZ-NRW GmbH, Aachen, Germany,

5. W. D. Heilmeyer, Technische Untersuchungen an römischen Grobronzen, Berichtsband **45** der DGZfP, 4. Int. Konf. ZfP an Kunst- und Kulturgütern, Berlin, 3.-8. Okt. 1994, vol.**1**, (1995) 11–20.
6. R.-B. Wartke, Zur Herstellung zweier Kupferstatuetten aus Assur, in Handwerk und Technologie im alten Orient: Ein Beitrag zur Geschichte des Technick im Altertum, Mainz (1994) 127–151,
7. catalog from the exebition: "Discoveries at Ashur on the Tigris, Assyrian Origins", ed. by P. O. Harper et al., The Metropolitan Museum of Art, New York, may 2-august 13, (1995) 38–39
8. Kataloge des Rheinischen Landesmuseums Bonn, Band 1,1: "Das Wrack. Der antike Schiffsfund von Mahdia", Rheinisches Landesmuseum Bonn, Landschaftsverband Rheinland, ed. by Hartwig Lüdtke, (1994), Rheinland-Verlag GmbH, Köln.
9. D. Wildung, Königsideologie und Computergrafie – Untersuchungen am Porträtkopf der Königin Teje, Berichtsband 45 der DGZfP, 4. Int. Konf. ZfP an Kunst- und Kulturgütern, Berlin, 3.-8. Okt. 1994, vol.**1**, (1995) 66–74.
10. M. Müller-Karpe, Gutechniken des dritten Jahrtausends in Mesopotamien, Rencontre Internationale Assyriologique, Prag, (1996), to be published.
11. D. Wildung: Metarmorphosen einer Königin, Antike Welt, vol.**26**(4), (1995) 245–249.
12. A. Hoffmann, J. Goebbels, B. Illerhaus, Virtuelle Rekonstruction des Schädels von Le Moustier - Projektvorschlag, Berichtsband 45 der DGZfP, 4. Int. Konf. ZfP an Kunst- und Kulturgütern, Berlin, 3.-8. Okt. 1994, vol.**1**, (1995) 408–411.
13. A. Flisch, T. Lüthi: CT Applications in Archeological and Paleontological / Anthropological Research, Industrial Computed Tomography Topical Conference, May 13-15, (1996), Huntsville, Alabama, USA.
14. T. H. M. Falke, An Egyptian mummy in the Bybels Museum, Amsterdam, Jaarbericht Ex Oriente Lux, vol.**26**, (1980) 35–38
15. M. Egg R. Goedecker-Ciolek, W. Groenman-van Waateringe, K. Spindler: Die Gletschermumie vom Ende der Steinzeit aus den Ötztaler Alpen, Jahrbuch des Römisch-Germanischen Zentralmuseums, Mainz, vol.**39**, (1992)
16. B. Hering: Die Untersuchung von Sammlungsgegenständen mit Hilfe der Computertomographie - Fallbeispiele, Berichtsband 45 der DGZfP, 4. Int. Konf. ZfP an Kunst- und Kulturgütern, Berlin, 3.-8. Okt. 1994, vol.**1**, (1995) 50–59
17. H. Born, St. Gumann: Grundlagenforschung Tauschierungen "Zur Herstellungstechnik und Restaurierung von Riemenzungen des 7. Jahrhunderts, Arbeitsblätter für Restauratoren, vol.**2**, (1992) 273–280.
18. J. Maish, J. P. Getty Museum, (1995), private communication.
19. H. P. Roger, Diplomarbeit, Fachhochschule Hildesheim / Holzminden, Germany.
20. B. Illerhaus, Fortschritte in der Computertomographie, Restauro, **101**(5), (1995) 344–349

Layered Artefacts: Non-destructive Characterization by PIXE and RBS

C. Neelmeijer, M. Mäder, W. Wagner
Research Center Rossendorf Inc., P.O.B. 510119, D-01314 Dresden, Germany
E-mail: neelmei@fz-rossendorf.de

H.-P. Schramm
Academy of Fine Arts, D-01288 Dresden, Germany

Abstract. To identify paint layer arrangements and their elemental composition without sampling, the ion beam techniques PIXE/RBS are successfully applied on air.

1 Introduction

Art objects are unique and often "sick patients". Hence, restoration and scientific work have to take place extremely gently. For example, contactless radiation area-examinations are applied to image underdrawings (Infrared radiation) or to investigate the white lead pattern (X-radiation) of paintings. Substantial analysis [1], however, is traditionally coupled to the preparation of a tiny specimen. This, on the one hand, represents a very crude action. On the other hand, it is questionable whether the sample represents the reality of the artefact. Perhaps the chosen position was formerly cemented or retouched. Regarding completely intact objects or delicate details, e.g. miniatures, substantial analysis via sampling is absolutely forbidden.

In connection with an external proton beam, a proton beam on air, the well known ion beam techniques PIXE (Proton Induced X-ray Emission) and RBS (Rutherford Backscattering) allow a multiple point analysis of chemical elements directly on the object and in a non-destructive manner [2,3]. Recently, the power of the method was increased using both PIXE with proton energy variation (δE-PIXE) and simultaneous PIXE/RBS measurements [4-7]. In this new regime not only the elemental components of pigments but also paint layer arrangements can be identified; the latter possibly allows to deduce the paint technique of a painter (layered composition or colours admixed on the palette).

2 Development of the Method

Sets of paint layer test structures have been used to develop the method. For example, typical historic pigments (e.g. verdigris, white lead) or modern paint materials (e.g. chromium oxide, zinc white) were put on a chalk ground in both possible sequences of layers, which differ in type and order of the pigments, as well as pigment admixtures. As tested, thicknesses of extended paint layers (~ 10...200 μm) can be estimated by means of δE-PIXE. Near surface layers in the depth dimension of only some microns (e.g. varnish coverings) are well characterized by the complementary RBS. For checking the PIXE/RBS results, cross sections were prepared from tiny samples and the individual film thicknesses were determined by optical microscopy.

3 Example of Application

The technique has been successfully applied to compare pigment components and layer arrangements of Lucas Cranach the Elder's paintings "14 Nothelfer" (earlier work) and "Christus und Maria" (later work) [8]. Very thin layers of book paintings were characterized mainly by assistance of RBS [9].

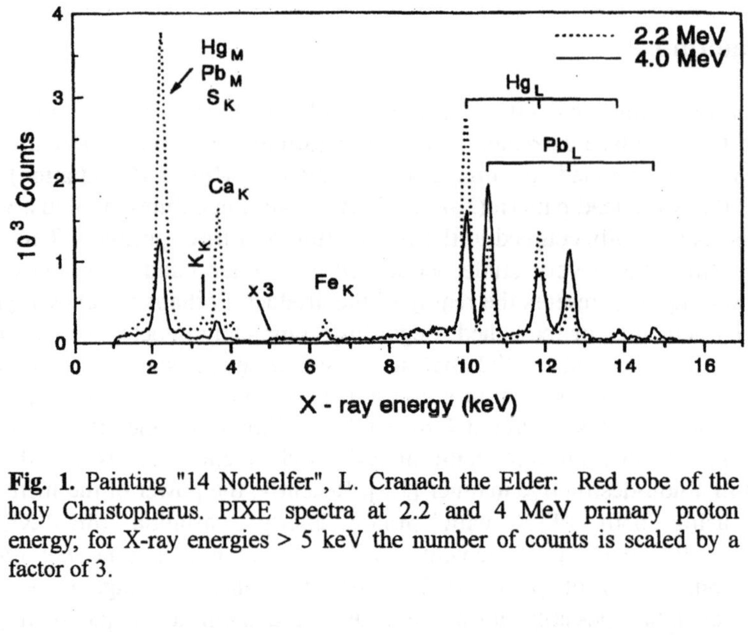

Fig. 1. Painting "14 Nothelfer", L. Cranach the Elder: Red robe of the holy Christopherus. PIXE spectra at 2.2 and 4 MeV primary proton energy; for X-ray energies > 5 keV the number of counts is scaled by a factor of 3.

Considering the relative variation of the characteristic X-radiation (Hg_L, Pb_L), emitted from the chromophoric elements Hg and Pb of cinnabar and white lead, respectively, it is remarkable that the intensity ratio Hg_L/Pb_L strongly decreases with increasing proton energy. This fact identifies the colours to be arranged in layers: cinnabar on white lead. Note that the S_K line cannot be resolved from the interfering Hg_M radiation. The Ca_K and Fe_K radiation results from impurities in the paint materials.

In analogy to the paint layers, gold jewelry has been proofed with respect to the depth dependence of the alloy composition making use mainly of external PIXE with proton energy variation [10].

This work is supported by BMBF 215-4003 NE9ROS.

References

1 H. P. Schramm and B. Hering, *Historische Malmaterialien und ihre Identifizierung*, (ADVA, Berlin und Graz, 1989).

2 M. Bernasconi, R. Cambria, P. Canart, L. Dal Poz, P. del Carmine, M. Grange, F. Lucarelli, J.D. MacArthur, P.A. Mandò, M. Maniaci and P. Sammuri, *Ancient and Medieval Book Materials and Techniques*, Studi e Testi II, p. 357 (1993), ed. by M. Maniaci and P.F. Munafò (Città del Vaticano, Biblioteka Apostolica Vaticana).

3 C. Neelmeijer, W. Wagner and H.P. Schramm, *Analysis of art objects by the combined application of three ion beam methods on air,* Proceedings of the 4th Int. Conf. on Non-destructive Testing of Works of Art, p. 296, ed. by DGZfp e.V. (Berlin, 1994).

4 C. Neelmeijer, W. Wagner and H.P. Schramm, Naturwissenschaften **81**, 553 (1994).

5 P.A. Mando, Nuclear Instruments and Methods **B85**, 815 (1994).

6 W. Wagner, C. Neelmeijer, Fresenius Journal of Analytical Chemistry **353**, 297 (1995).

7 W. Wagner, C. Neelmeijer and H.P. Schramm, *Paint layer studies using PIXE and RBS on air,* Proceedings of the 4th Int. Conf. on Non-destructive Testing of Works of Art, p.316 , ed. by DGZfp e.V. (Berlin, 1994).

8 C. Neelmeijer, W. Wagner and H.P. Schramm, *Depth Resolved Ion Beam Analysis on Art objects,* Proceedings of the 12th Int. Conf. on Ion Beam Analysis (Tempe, Arizona, 1995), Nuclear Instruments and Methods **B**, in press.

9 C. Neelmeijer, W. Wagner, H.P, Schramm and U. Thiel, Nuclear Instruments and Methods **B99**, 390 (1995).

10 G. Demortier and J.L. Ruvalcaba-Sil, *Differential PIXE analysis of Mesoamerican jewelry items, Proceedings of the* 12th Int. Conf. on Ion Beam Analysis; (Tempe, Arizona, 1995), Nuclear Instruments and Methods **B**, in press.q

New Methods of Reflectography with Special Filter and Image Processing Techniques: Examination of Materials, Writings and Underdrawings

Robert Fuchs, Ralf Mrusek
Fachhochschule Köln, Restaurierung und Konservierung von Schriftgut, Graphik und Buchmalerei,
Ubierring 40, D-50678 Köln, Germany
E-mail: fuchs@re.fh-koeln.de, mrusek@re.fh-koeln.de
Internet: www.fh-koeln.de//presse/forsch/methoden/forsch11.html

Abstract. Investigations of old manuscripts and documents should be made totally non-destructive. New IR/UV/Vis-reflectography techniques, combinations of bandpass filters with image processing, allow one to distinguish inks and colour materials and can make faded or rubbed writings and underdrawings visible.

The illumination of an object with light has two different effects: absorption and remission. Therefore, in the visible range (Vis) the various colors can be observed. The relationship between the intensity of light and its wavelength can be quantified with spectrometers (UV, Vis, IR). Using reflectography we take advantage of the fact that each material responds to light in a characteristic manner.

This method can be applied to analyze materials of a painting or to distinguish between inks or other drawing materials[1]. It is useful to make visible underdrawings in book illumination[2] or to examine faded texts[3]. The analysis includes the range of

[1] For the non-destructive (i.e. without any samples) analysis of colour and ink materials three other techniques are already in use: XR-diffractometry, IR-spectroscopy and colour spectroscopy, see: *Robert Fuchs*, Zerstörungsfreie Untersuchungen an mittelalterlicher Buchmalerei - eine wissenschaftliche Herausforderung. Tagungsband zum Symposium für Zerstörungsfreie Prüfung von Kunstwerken, 19./20. November 1987. Deutsche Gesellschaft für Zerstörungsfreie Prüfung e.V. Berichtsband 13, Berlin 1988, pp. 120-127. Other techniques need samples from the original, which is very problematic from the conservational point of view: for example the RAMAN-spectroscopy, see: *Bernard Guineau, Jean Vezin*, Etude technique des peintures du manuscrit De laudibus sanctae crucis conservé à la Bibliothèque Municipale d'Amiens (Amiens 223). Scriptorium 46.1, Bruxelles 1992, pp. 224-237.
The new reflectography technique allows the non-destructive 2-dimensional material analysis: *Robert Fuchs, Ralf Mrusek, Doris Oltrogge*: Spektrale Fenster zur Vergangenheit. - Ein neues Reflektographieverfahren zur Untersuchung von Buchmalerei und historischem Schriftgut, Naturwissenschaften 82.2, Heidelberg 1995, pp. 68 -79, esp. p. 73, fig. 1, 2

[2] *Robert Fuchs, Doris Oltrogge*, Naturwissenschaft und Stilkritik - Handschriften aus dem Umkreis des Registrum-Meisters. In: Kunsthistoriker, Mitteilungen des Österreichischen

the ultraviolet (UV), the visible and the infrared (IR). In all cases, problems can be solved by analyzing separate sections of the spectrum. To realize such a segmentation of the spectrum, bandpass filters can be applied (fig. 1).

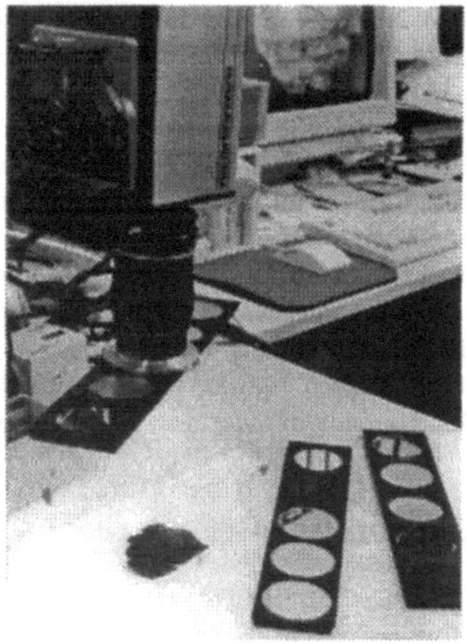

Fig. 1. Under the IR-camera lies a burnt dark birchbark-manuscript. Under the objective of the camera a set of bandpass-filters is mounted. In the background on the computer screen the text of the manuscript can be read.

1 Method

The object is illuminated with light of low intensity. Depending on the problem we use various light sources, which give access to the different parts of the spectrum. A

Kunsthistorikerverbandes, Jg. VIII (Sondernummer). Wien 1991, pp. 96-104. *Doris Oltrogge*, "Materia" und "ingenium". Beobachtungen zur Herstellung des Egbert-Codex. Egbert - Erzbischof von Trier. Hrg. v. F. J. Ronig unter Mitarbeit von A. Weiner und R. Heyen. [=Trierer Zeitschrift, Beiheft 18], Trier 1993, Vol. 2, pp. 123-52. *Robert Fuchs, Doris Oltrogge, Ralf Mrusek*: Eine Galerie des Unsichtbaren. Spektrum der Wissenschaft, Juni 1995, pp. 85 - 89. *Robert Fuchs, Doris Oltrogge*, Neue Untersuchungen an mittelalterlichen Handschriften. In: Rhythmus und Saisonalität. Akten des 5. Symposiums des Mediävistenverbandes. Hrg. P. Dilg, G. Keil, D.-R. Moser. Sigmaringen 1995, pp. 327 - 345.

[3] *Robert Fuchs, Ralf Mrusek, Doris Oltrogge*, Die Entstehung der Handschrift - Materialien und Maltechnik. In: Petrus von Ebulo, Liber ad honorem Augusti sive de rebus Siculis, Eine Bilderchronik der Stauferzeit aus der Burgerbibliothek Bern. (Ed. Th. Kölzer und M. Stähli), Sigmaringen 1994, pp. 275 - 285. *Ralf Mrusek*, Reflections reveal faded secrets of ancient books. Opto & Laser Europe 36, Bristol 1996, p. 11.

The filter technique is also a powerful tool in the examination of underdrawings. Sometimes there are considerable differences between the drawing and the completed painting[11]. Whether the drawing materials are recognizable or not on the digital image depends on the property of the materials covering the drawings. We found that in a range where distinct drawing materials are observable, some painting materials will become transparent but others are impenetrable to the light in the chosen bandpass region. Therefore, a series of images recorded with different bandpass filters have proved useful to obtain mostly complete information on the underdrawings. In some cases, different drawing materials have been used in one picture. The use of bandpass filters allows to concentrate on one material to make it visible and to suppress the effect of the other ones in the digital image.

3 Image Processing

The interpretation of images of ancient objects in art historical research can be supported by a digital image processing system[12]. Thus, images produced by IR cameras and optical filtering will lead to a better understanding of the manufacturing process and the history of an object, but the method has some disadvantages due to the technique. Therefore these disadvantages have to be corrected by developing the necessary software programmes[13].

The available Image Processing software can be used to optimize the image acquiring procedure and to improve the stored images in additional processing steps[14].

3.1 Image Acquiring

In digital image acquisition, noise effects which interfere with the signal are introduced by the electronic components of the system. They do have a recognizable effect, because we are working with weak signals. For illumination, one has to be careful in choosing the light source. Light of lower intensity should be applied to avoid heating up fragile surfaces. But the smaller the intensity of light for illumination, the smaller the response of light from the pigments and inks. That means that the light which comes to the camera is already limited in intensity. In addition, the optical filter process will cut only a narrow band of the whole light intensity. This weak signal which is reaching the detector has to be amplified

---

[11] *Fuchs et al.*, Eine Galerie (supra note 2), fig. 3, 4, 5.

[12] *Andreas Burmester, F. Bayerer*, Towards improved infrared reflectograms. Studies in Conservation 38.3, London 1993, pp. 145 - 152.

[13] We use for the acquisition of the data a IC-P frame grabber board (Stemmer, Puchheim) together with OPTIMAS 6.0. This software can be programmed by a computer language similar to C. For the image enhancement we use ADOBE PHOTOSHOP 2.5 and 3.5.

[14] *Peter Haberäcker*, Digitale Bildverarbeitung, München 1989 (3. Aufl.). *Bernd Jähne*, Digitale Bildverarbeitung, Heidelberg 1990.

electronically. But in the same way the signal is amplified, the noise of the image acquiring system is increased. Therefore, besides the signal the noise is recognizable in the resulting images which are fuzzy and blurred. It is desirable to suppress the noise interference in advance during the acquisition process by digital image processing.

Therefore we have implemented the following acquisition procedure. The electronic noise is changing in time, thus it can eliminated by storing some images in a time interval and carrying out an averaging process. A 16 Bit Frame can be opened to realize this process. A series of 8 Bit images will be copied to the frame, the images are added and the result is divided by the number of images to get a 8 Bit result.

Through this operation, the noise in the resulting image can be reduced in comparison to an image frozen immediately. To get an impression of the amount of noise reduction, we have measured it in the following way: we acquire images of the same object with altering numbers of images taken in the average process. We are able to take at most 128 images in one procedure. In this case, the noise reduction should be maximal. We perform a discrete Fourier transform[15] to quantify the amount of noise in the images. The noise recognizable in the images as a graytone variation of high frequency superposing the whole signal will be described in the Fourier spectrum in a more suitable manner. The spectrum of the noise does have portions mainly in the high frequency domain, while the main part of the signal will be found in the low frequency region. In fig 4 a Fourier spectrum of an image is shown which has been acquired without additional image processing. The spectrum of the same image acquired in 128 cycles is displayed in fig. 5.

Fig. 4. Fourier spectrum of an image acquired without averaging procedure.

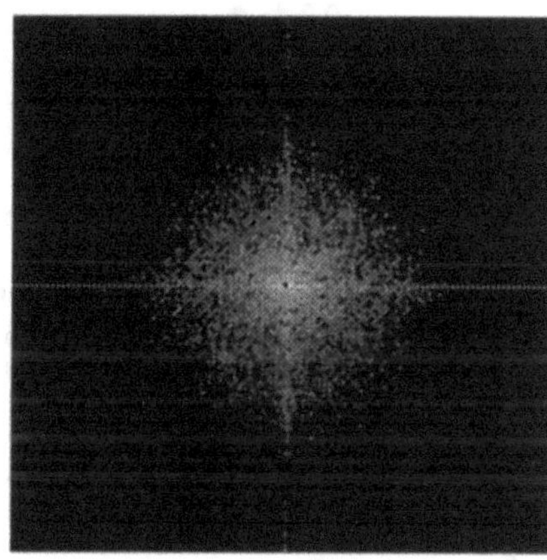

Fig. 5. Fourier spectrum of an image acquired with averaging in 128 cycles.

[15] *Elbert Oran Brigham,* FFT - Schnelle Fouriertransformation, München/Wien 1989.

the transition from a hill to a valley that the contrast is increased and the grayvalues are swinging up.

4. Conclusion

The different software image processing techniques together with the hardware system, bandpass filter, tunable monochromatic lighting system, and the three different cameras (UV, Vis, IR) allow us to get images of high quality in the range of 200 - 1900nm. These images give much information about the spectral characteristic of the materials. The interpretation of this information makes it possible to analyse works of art with a totally non-destructive and portable technique.

X-Ray Fluorescence Analysis Induced by Synchrotron Radiation (SYXRF) and First Archaeometric Applications *

Hans Mommsen, Heiko Dittmann, Anno Hein and Achim Rosenberg

Institut für Strahlen- und Kernphysik, Universität Bonn, Nußallee 14–16, D-53115 Bonn, Germany
E-mail: mommsen@iskp.uni-bonn.de

Abstract. The SYXRF method used at the storage ring ELSA in Bonn and its properties are found to be very well suited for elemental analysis in investigations of valuable objects pertaining to the history of art and civilization (metals, pigments, inks)

1 Introduction

X-ray fluorescence (XRF) analysis is a well established method and is often used in archaeometry and other work dealing with valuable objects pertaining to the history of art and civilization. The reason is that all different types of XRF using different excitation modes are non-destructive, so that no sample has to be taken from the objects. Rather new is the special fluorescence excitation mode by synchrotron radiation (SR) beams termed SYXRF. After introducing the basic principles of this method and the apparatus installed at the ELectron Storage Accelerator (ELSA) in Bonn, its properties, advantages and limits are described. As examples, first results are presented of archaeometric applications and projects concerning the analysis of icons and early printings.

2 SYXRF Method

The white spectrum of SR at the high energy storage ring ELSA is used for excitation of X-ray fluorescence (SYXRF). Besides the well known general properties of the XRF and PIXE (particle induced X-ray emission) methods (instrumental, multielemental, quantitative, non-destructive, surface analysis) the main advantages of SYXRF are

- high SR intensity, therefore
- small beam diameter down to a few microns possible (point analysis)
- fast (few minutes measuring time)

* This work has been funded by the German Federal Minister for Education, Research, Science and Technology (BMBF).

- low energy deposition in sample
- easy sample handling: no vacuum needed for elements with $Z > 14$
- independent of any standards (Fundamental Parameter Method (FPM))
- high sensitivity for trace elements ($\approx 1\,ng/cm^2$)

Quantitative determinations of concentrations ($\mu g/g$) in thick samples or areal densities (ng/cm^2) for surface layers are done by the FPM, i. e. a theoretical prediction of the fluorescence line intensities iterating the sample composition until agreement with the measured intensities is obtained[1]. For these calculations the energy distribution of the exciting SR has to be known and its attenuation in the sample. The strong variability of the SR beam intensity and spectral distribution even during one short measurement made an on-line beam diagnostics system[2] necessary to monitor these parameters. Furthermore, the fluorescence production and attenuation cross sections and any secondary excitations (enhancements) are considered using tabulated atomic data values. The precision of the quantitative calculations depends on the errors of these input parameters and is $< 3\,\%$–$7\,\%$. The different components of the apparatus and the correctness of the set of written programs was tested on a number of standards with certified concentration values.

3 Examples of first applications

Hot Gilded Layer of a Russian Metal Icon

The quantitative analysis of homogeneous metal objects not containing light elements (not visible in the X-ray spectra) is simple. A number of small metal icons very common in Russia was analysed and not, as commonly assumed, bronze, but brass was detected in most cases. A special case was the analysis of a hot gilded triptych. Since we are able to calculate fluorescent intensities absolutely and assuming, that the elements Au, Ag and Hg are in the gilded surface layer only and Cu and Zn lines in the X-ray spectrum stem from the underlying brass, the composition as well as the thickness of the gilded layer can be obtained additionally to the brass composition. The results for the triptych are given as example - gilded layer: 87% Au, 2.5% Ag, 10% Hg, average thickness 3 μm; underlying brass: 72.3% Cu, 24.3% Zn, traces of Pb, Ni, Sn; brass measured at a non-gilded position: 73.0% Cu, 26.2% Zn, same traces.

Painting Technique of a 'Tabletka' Icon

Tabletka icons painted on the front and backside are very rare and valuable. In collaboration with I. Bentchev, the paint layers and pigments of such an icon from the 16th century (uncovered under 5 layers of later paintings) were analysed. It is believed that it was used as a model with respect to iconography and the techniques of painting in the monk's workshops at Novgorod. In some places, small pieces of paint layers had fallen off exposing the underlying layers. With

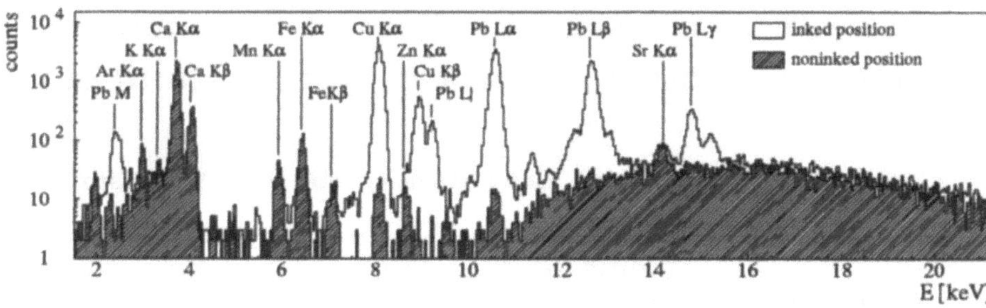

Figure 1. X-ray spectra of ink & paper and paper only (hatched) of the B-36 excited with white SR (2.3 GeV electron energy, 30 mA, 1.2 mm Al absorber, measuring time 300 s); result of ink analysis: Cu 32 μg/cm^2, Pb 54 μg/cm^2

the small SR beam diameter of about 100 μm, such steplike structures could be analysed. Besides the nature of the pigments in use questions concerning the sequence and the originality of some of the paint layers could be answered. From further such studies we hope that in the future we will be able to differentiate between different Russian workshops and production times.

Ink Analysis in Incunabula

After the first PIXE measurements in the USA[3] it was hoped that a characteristic elemental composition would be recorded in the dried inks of early printings, allowing the determination of unknown printing-offices or printers. In collaboration with M. Boghardt several incunabula, including the Biblia Latina B-36 and the Catholicon, both made by unknown printers, could be analysed. A spectrum from an inked position of a page of the B-36 is shown in Fig. 1. The results are not promising. Although information can be deduced about the printing procedure, i. e. about the sequence of printing the different recto and verso sides of a book seen by a change of the ink composition, the second goal, the identification of typical ink compositions used by different early printers seems difficult, considering the small number of measurable elements and the variable concentrations observed. The final implications of the many data taken meanwhile have yet to be investigated and discussed with experts of early printing.

References

1 F. J. Pantenburg, Th. Beier, F. Hennrich and H. Mommsen, Nucl. Inst. Meth. **B68**, 125 (1992)
2 D. Heimermann, Th. Beier, H. Dittmann, A. Hein, H. Mommsen and A. Rosenberg, Naturwissenschaften **81**, 551 (1994)
3 T. A. Cahill, B. Kusko and R. N. Schwab, Nucl. Instr. Meth. **181**, 205 (1981)

Neutron Activation Analysis of Mycenaean and Related Pottery from the Greek Mainland *

Joseph Maran[1], Anno Hein[2], Doris Ittameier[1] and Hans Mommsen[2]

[1] Institut für Ur- und Frühgeschichte, Universität Heidelberg, Marstallhof 4, D-69117 Heidelberg, Germany
[2] Institut für Strahlen- und Kernphysik, Universität Bonn, Nußallee 14–16, D-53115 Bonn, Germany
E-mail: mommsen@iskp.uni-bonn.de

Abstract. Neutron Activation Analysis (NAA) is applied to determine different production series and the places of the corresponding potter's workshops of Mycenaean ceramics by their chemical patterns. Until now, more than 1500 samples have been analysed and about 80 different patterns were detected, but, at present, only 6 of them can be assigned to specific regions of production. However, first archaeological results concerning e. g. amphoroid craters with chariot scenes have been obtained.

1 Introduction

The fine decorated pottery of the Greek Mainland in the Late Bronze Age (the Mycenaean period (1600–1050 B. C.)) has been a focus of archaeological interest for more than a 100 years, and this for several reasons. Not only does this pottery have a wide distribution, encompassing almost the whole Mediterranean, but it also shows such a well defined range of vessel shapes and motives that even smallest fragments can be attributed with certainty to specific vase types. Considering its wide distribution and amazing homogeneity, several questions immediately arise: 1. To what degree was this kind of pottery manufactured in only a few places, or rather on a decentralized basis in the different regions of its distribution? 2. Was there some sort of internal ranking between production places of only regional, and those of supra-regional importance? Moreover, if this were the case, would it be possible that the most important workshops were linked to Mycenaean palaces?

2 Analytical Procedure and Sample Choice

The principles of provenancing pottery by 'chemical fingerprinting' are well known. All sherds belonging to the same production series made by a given fixed 'recipe' from local clays or clay mixtures will show the same concentration

* This work has been funded by the German Federal Minister for Education, Research, Science and Technology (BMBF).

pattern of elements. If the patterns of a pottery production place are known from measurements of reference material (i. e. wasters from potter's kilns), sherds with similar patterns can be assumed to originate from that place. Neutron Activation Analysis (NAA) is especially well suited for provenancing of pottery, since it fulfills the necessary demands very well. We are able to measure more than 25 elements down to the trace element level with precisions of a few percent. The grouping of samples of similar chemical composition is done by a novel filter method[1] using a specially adapted similarity measure developed in Bonn, which is a modified Mahalanobis distance including consideration of experimental errors and possible constant shifts of data due to pottery making practices.

Our project started in the mid-1980's, and up to now more than 1500 samples have been analysed. The analysis of more than 1300 new samples from different regions of Mainland Greece (Fig. 1a) taken recently is still underway.

3 Provenance of Chemical Patterns

As mentioned, the definition of reference groups for distinct pottery production places by the analysis of wasters from potter's kilns is crucial for our research. This, however, for several reasons has proved to be a difficult task. Not only are there very few Mycenaean potter's kilns on the Greek Mainland, but these kilns only rarely yielded misfired pieces, clearly identifiable as refuse. Thus, although the number of different chemical patterns detected already now exceeds 80, only half a dozen can be assigned with a high probability to specific regions of production (Fig. 1b). In the Argolid, two main patterns are linked with specific locations. The first pattern, called Mycenae/Berbati (MB) is connected through Perlman's analysis of wasters to the potter's quarter at Berbati in the hinterland of Mycenae; the second one, called Tiryns/Asine (TA), through our analysis of wasters to a workshop in the region of Tiryns[2]. Especially the pattern MB proves to be dominant not only in the Mycenaean pottery from the Argolid, but throughout Greece and also in the one which we analysed from Egypt.

The recent analysis of samples from two amphoroid craters with chariot scenes from the Metropolitan Museum in New York offers further clues as to the significance of the MB pattern[3]. Both vessels were found on Cyprus, the first one at Maroni, the second one at Ayia Paraskevi. For years the question of the place of origin of these amphoroid craters has been a highly controversial issue. In spite of a distribution pointing to a Cypriot or even Near Eastern origin, there is growing evidence that these amphoroid craters of the 14th and early 13th century B. C. were produced in the Argolid. Thus, around 50 fragments of chariot craters were found at Berbati, some of which can be identified as refuse from a potter's kiln[4]. Our analysis clearly supports the notion of an Argive origin of these high quality vases, because both craters belong to the MB group. Accordingly, as A. Akerström has already pointed out, an unexpected picture emerges in which a workshop in the vicinity of Mycenae produced pottery for export to Cyprus and the Levant, where these Chariot craters evidently were much more popular than on the Greek Mainland.

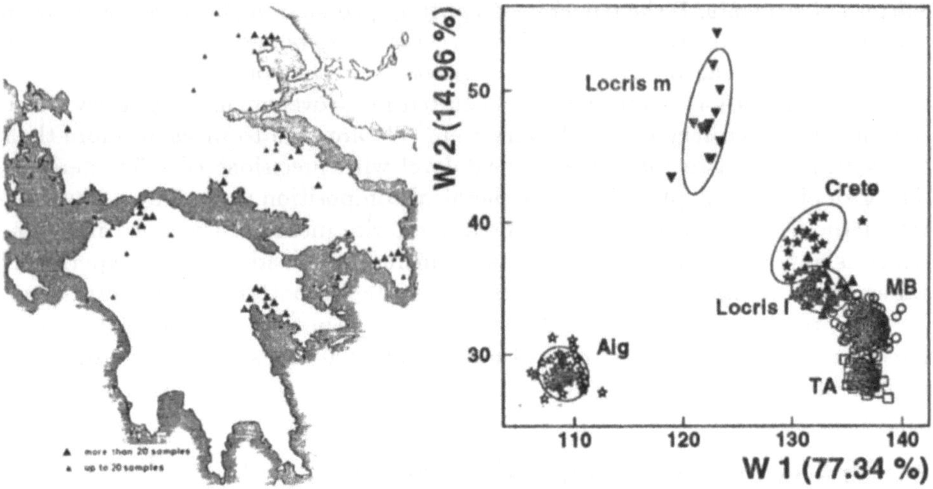

Fig. 1. (a) Bronze Age locations sampled in Greece (b) Discriminant analysis of 432 samples assuming 6 groups (MB = Mycenae/Berbati, TA = Tiryns/Asine, Aig = Aigina). The ellipses drawn are the 2σ boundaries of the groups.

These results underline the potential of NAA to add new insights to archaeological problems. A characterization of more regional workshops producing Mycenaean pottery would contribute significantly to a better understanding of the mechanisms underlying pottery production and distribution during the Bronze Age in Greece and the surrounding regions.

Acknowledgement

We owe a very special debt of thanks to the Greek Ministry of Culture and the numerous Greek colleagues, who have supported our project. We would also like to thank Dr. J. R. Mertens (Metropolitan Museum, New York) for giving us the opportunity to sample the two craters. The help of the staff of the reactor in Geesthacht is thankfully acknowledged.

References

1. Th. Beier and H. Mommsen, Archaeometry **36**, 287 (1994)
2. H. Mommsen, E. Lewandowski, J. Weber and Ch. Podzuweit, in: R. G. V. Hancock and L. A. Pavlish (eds.), *Proceedings of the 26th Int. Archaeometry Symp.*, Toronto, Canada (University of Toronto Press, Toronto 1988) 165
3. E. Vermeule and V. Karageorghis, *Mycenaean Pictorial Vase Painting* (Havard University Press, Cambridge [Mass.] and London, 1982) 196 No III.16; 200 No V.2
4. A. Akerström, *Berbati 2. The Pictorial Pottery* (Aströms Förlag, Stockholm, 1987) 26

Luminescence Dating in Archaeology: Recent Developments

U. Rieser*, A. Lang*, G. A. Wagner*, M. R. Krbetschek$ and W. Stolz§

* Forschungsstelle Archäometrie der Heidelberger Akademie der Wissenschaften am Max-Planck-Institut für Kernphysik, Postfach 10 39 80, D-69029 Heidelberg, Germany, E-mail: gwagner@goanna.mpi-hd.mpg.de
$ Forschungsstelle Geochronologie Quartär der Sächsischen Akademie der Wissenschaften zu Leipzig, Bernhard-von-Cotta-Str. 4, D-09596 Freiberg, Germany, E-mail: tl-labor@orion.hrz.tu-freiberg.de
§ TU Bergakademie Freiberg, Inst. f. Angew. Physik, Bernhard-von-Cotta-Str. 4, D-09596 Freiberg, Germany

Abstract. Luminescence dating is known as a reliable technique for age determination on ceramics and quaternary sediments. Based on recent technical developments, the method now reaches a substantially extended field of applications. A survey of the fundamental physical research will be given as well as case studies of new techniques like 'OSL subtraction dating' or 'easy-to-bleach-peak dating'.

1 Introduction

The luminescence techniques have evolved over the last 40 years to a powerful dating instrument in archaeology. Depending on how the luminescence is stimulated, one distinguishes the phenomena of *thermoluminescence (TL)*, *optically stimulated luminescence (OSL)* and *infrared stimulated luminescence (IRSL)*. Each of these phenomena has its specific potential for dating various archaeological materials in the time range from the medieval back to the palaeolithic periods. The OSL and IRSL techniques are sometimes treated together as "optical dating". The luminescence techniques differ from other major dating techniques, such as ^{14}C, essentially by their applicability to inorganic materials, their wide age-range from about 100 years to more than 100,000 years and the kind of datable events which are the last exposure to heat or to light.

Thermoluminescence dating was originally proposed in 1953 by Daniels et al. [1]. Ceramics and bricks were the first archaeological materials on which the new method was tested [2]. In the following years, thermoluminescence became a standard dating method for a variety of burned materials, mainly developed by Martin Aitken's group at Oxford. Starting with the recognition by Wintle and Huntley in 1979 [3] that solar bleaching resets the luminescence signal,

during the resetting and the stimulation processes. Investigations showed that just a few minutes of exposure to daylight erase the OSL signal. This rapid bleaching also allows the dating of such events when objects were only weakly or briefly illuminated. Therefore, a variety of samples like waterborne or archaeological sediments become datable, which are undatable with TL. Also very young sediments, let's say less than several thousand years, can be dated due to the missing residual. Physical studies have shown that the OSL and TL phenomena are largely independent from each other [10]. Apparently they are based on different populations of electron traps.

The optimum stimulation wavelength depends on the type of mineral. Depending on the colour of the stimulating light, one subdivides OSL in several types; in practice these are the infrared stimulated luminescence (IRSL) used for feldspars and green light stimulated luminescence (GLSL) for quartz. In the case of finegrained polymineral samples, the selection of the stimulation wavelength allows, in principle, to probe selectively the feldspar or the quartz fractions.

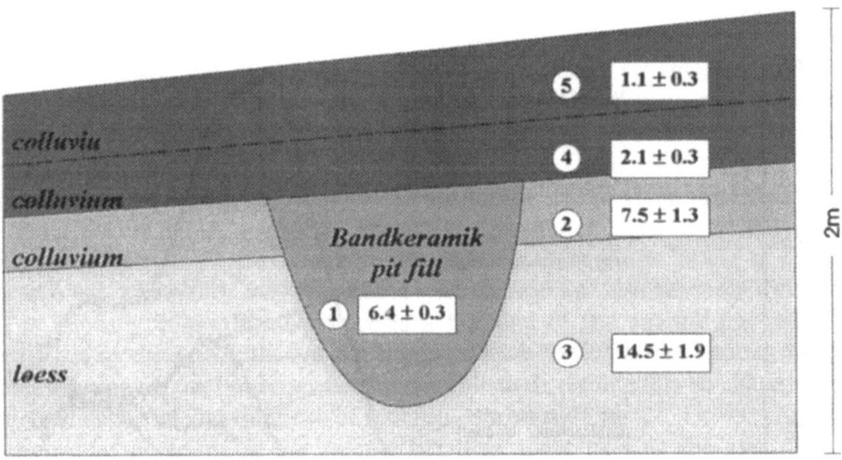

Fig. 2. Schematic cross-section of the 'Bandkeramik' site Bruchsal-Aue/Germany with sample locations and IRSL-ages (in thousand years).

As an example Lang and Wagner [11] were able to date the colluvia filling of an Early Neolithic pit (Bandkeramik) properly with 6.4 ± 0.8 ka. This result was surprising, because due to the topographic location, the transportation distance of the sediments can not have exceeded 100 m, which implicates very short time for light exposure and, thus, imperfect luminescence signal resetting.

During OSL measurement, a sufficiently large signal can be obtained without destroying the whole signal. This enables the determination of the archaeo-dose on few or even on single sample-aliquots [12]. Conventionally, 40 to 80 aliquots are required, and since these aliquots rarely are equal, some technique of normalization needs to be applied. This implies large sample size, tedious and time-consuming experimental procedures as well as low precision. Therefore, a lot of effort is presently invested into the development of single aliquot and single grain OSL dating.

4 Research on the physical basis of luminescence dating

As already shown, there are some amazing applications for the recently developed luminescence techniques, and the potential is still large. The crucial point is the limited knowledge about the physical system underlying important characteristics, such as stability and bleachability, of the luminescence signal. One way to get more information is to analyse the emission spectra of the minerals [13]. Experimentally this is not an easy task because of the very low light levels we have to deal with, in fact, counting of single photons is required. Also, every trap can be emptied only once, giving a fixed upper limit of overall detectable light which is not reproducible. Since these few photons have to be dispersed on a multichannel detector by a spectral analysing system, one needs highest-sensitive and lowest-noise instrumentation.

In 1992, a joint project of our laboratories at Heidelberg and Freiberg started to build such an apparatus [14]. A liquid nitrogen cooled CCD-camera with special read out electronics was used as a detector. It is possible to obtain thermoluminescence spectra of even weak emitting samples due to the properties that (i) the dark signal is less than one electron per pixel per hour and (ii) the read out

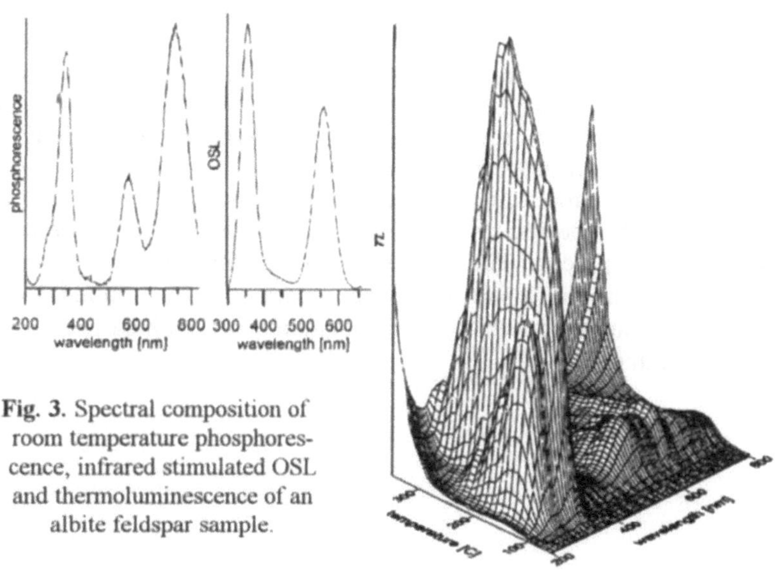

Fig. 3. Spectral composition of room temperature phosphorescence, infrared stimulated OSL and thermoluminescence of an albite feldspar sample.

noise is only about 5 electrons per channel.. The temperature range is -150 to 700 °C, and the spectral region reaches from 200 to 800 nm. It is worth mentioning that this was worldwide the first spectrometer capable of taking IRSL-spectra of feldspars, and still is the most sensitive one.

Using this equipment, a set of 30 feldspars of different composition well characterized by x-ray diffraction and atomic emission spectrometry was investigated. Spectra of thermoluminescence, room temperature phosphorescence and infrared stimulated luminescence were analysed. It was shown that five main emissions in the bands around 280, 330, 410, 560 and 700 nm exist [15].

The centre wavelength of the peaks changes slightly depending on the measurement method and type of feldspar. In the case of the 700 nm emission, a proportionality between potassium content and peak wavelength was observed [16]. This will lead us to a new and very simple method of ^{40}K determination, one of the three contributors to the natural dose-rate.

Experiments regarding bleachability and thermal stability of the main emissions were undertaken. It was shown that the 560 nm line is especially useful for young samples up to about 50.000 years because of its rapid bleachability. On the other hand, the 410 nm peak is suitable for older samples because of its high stability. Surprisingly, the blue luminescence is not completely reset when a large fraction of ultraviolet light is present during beaching [17]. This is, for example, the case in sediments deposited in clearwater near the seashore or in rivers, and must be considered as a possible source of age overestimation.

The 280 nm emission was shown to be thermally instable, and therefore has to be removed from the detected signal for luminescence dating.

For a better understanding of the luminescence mechanism, it would be helpful to know more about the type and structure of the emission centres. The above mentioned set of 30 feldspars was quantitatively analysed regarding the elements Ag, B, Ba, Cu, Ga, Mn, Pb, Sn, V, Be, Co, Cr, Mo, Nb, Ni, Ti, W, Y, Zr. A correlation between composition and emission spectra could not be established, which indicates the predominate importance of structural parameters.

The model of the IRSL process, which had been developed in 1988 by G. Hütt, was tested and modified by spectral investigations. In principle, the model describes IRSL as a thermo-optical process. The electron is evicted from its trap by infrared light but does not reach the conduction band. Instead, it is captured by an intermediate trap where again it needs some stimulation energy, which is delivered by the ambient temperature.

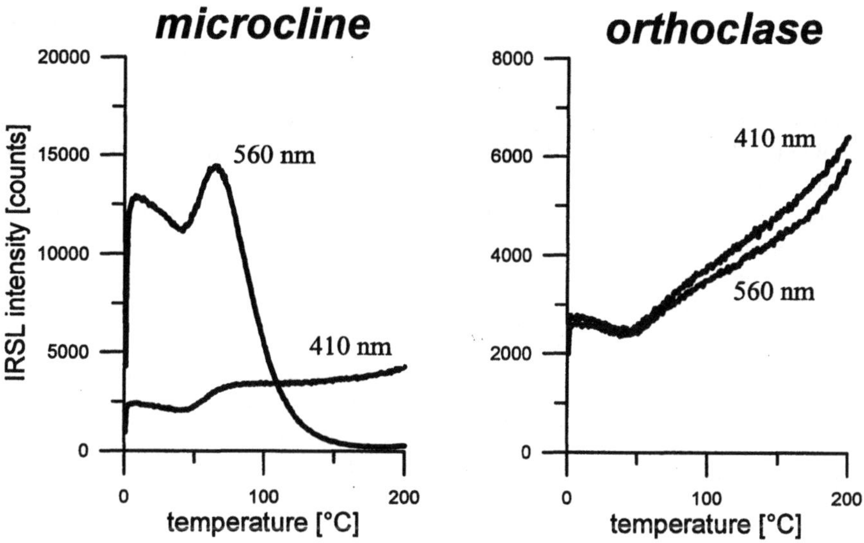

Fig. 4. IRSL of microcline and orthoclase feldspar, measured during a heating ramp of 4 K/s up to 200 °C. The 560 nm emission of microcline drops in intensity above 70°C.

To substantiate this model further, the luminescence of a microcline extracted from a sediment sample was measured using the spectrometer during infrared stimulation at different temperatures between -150°C and +200°C.

In accordance with the model, the intensity of the 410 nm emission rose uniformly with temperature. But the 560 nm IRSL dropped above 70°C down to approximately zero, in contradiction to G. Hütt's model, as shown in Fig. 4.

Tests with other feldspars showed that this behaviour seems to be typical, especially for microclines. Orthoclase, which has the same chemical composition as microcline but has a monoclinic lattice (instead of a triclinic one for microcline), behaves as prescribed by the model. Another interesting feature of microcline is the rise of the 560 nm luminescence emission during the first 500 milliseconds of stimulation. As an explanation for both observations, the existence of a metastable state in the 560 nm recombination path was postulated, which can be emptied thermally above 70°C. This state can be interpreted as a major sourcc of crror when using conventional luminescence dating procedures [18]. Techniques to circumvent this problem will be developed in the near future.

5 New techniques and applications

Influenced by the spectral studies, A. Lang and G. A. Wagner developed a new procedure for IRSL dating of poorly bleached finegrained sediments [19]. They used a narrowband optical filter to exactly detect the 410 nm luminescence emission band.

In addition, the subtraction dating technique was applied [20]. This technique utilizes the 'hard-to-bleach-component' in the measured signal, which is quantifiable using the late (after one minute of stimulation) luminescence light. The late signal is subtracted from the early signal consisting of a mixture of easy- and hard-to-bleach-components. The remaining difference gives access to extremely light sensitive portions of the stored age information.

This method is now successfully used mainly in reconstructing agricultural history, which is reflected by colluvial sediments accumulated in river valleys. This enables the indirect dating of the countless artifacts in these sediments.

The IRSL method, as mentioned before, requires only little sample-material and inherently has a higher precision compared to traditional thermoluminescence dating. It is obvious to also apply this technique to ceramics, where these advantages are required in particular. First successful attempts were made in Heidelberg using Precolumbian artifacts from Ecuador. For ceramics, the newly developed optical dating might replace TL-techniques on a long-term time scale.

6 Conclusions and perspectives

Luminescence has become a versatile dating instrument in archaeology and Quaternary geology. It is applicable to various inorganic archaeological materials, as well as to geological events related to archaeology. Originally a ceramic-only dating method with, admittedly, limited accuracy, it has lately evolved into a unique tool for Palaeolithic chronology beyond radiocarbon and, even more exciting, for geoarchaeological reconstruction of prehistoric man's palaeoenvironment.

Acknowledgements. We gratefully acknowledge the support of our current work within the scope of the project 'New Technologies in the Arts', sponsored by the Bundesministerium für Bildung, Forschung, Wissenschaft und Technologie der Bundesrepublik Deutschland (BMBF).

References

1 F. Daniels, C. A. Boyd, and D. F. Saunders, Science **117**, 343 (1953).
2 N. Grögler, F. G. Houtermans, and H. Stauffer, Convegno sulle datazioni con metodi nucleari Proceedings, 5 (1958).
3 A. G. Wintle and D. J. Huntley, Nature **279**, 710 (1979).
4 D. J. Huntley, D. I. Godfrey-Smith, and M. L. W. Thewalt, Nature **313**, 105 (1985).
5 G. Hütt, I. Jaek, and J. Tchonka, Quatern. Sci. Rev. **7**, 381 (1988).
6 M. J. Aitken, *Science-based Dating in Archaeology* (Longman; London, 1990).

7 G. A. Wagner, *Altersbestimmung von jungen Gesteinen und Artefakten* (Enke; Stuttgart, 1995).
8 J. R. Prescott and P. J. Fox, Ancient TL **8**, 32 (1990).
9 P. Karelin, Dissertation, Univ. of Heidelberg (1997, in preparation).
10 H. Wiggenhorn, A. Lang, and G. A. Wagner, Naturwissenschaften **81**, 556, (1994).
11 A. Lang and G. A. Wagner, Archaeometry **38**, 129 (1996).
12 G. A. T. Duller, Radiat. Meas. **24**, 217 (1995).
13 P. D. Townsend and Y. Kirsh, Contemp. Phys. **30**, 337 (1989).
14 U. Rieser, M. R. Krbetschek, and W. Stolz, Radiat. Meas. **23**, 523 (1994).
15 M. R. Krbetschek and U. Rieser, Radiat. Meas. **24**, 473 (1995).
16 C. Dütsch, M. R. Krbetschek, D. Baake, U. Rieser, D. Wolf, and W. Stolz, Chemie der Erde (in preparation)
17 M. R. Krbetschek, U. Rieser and W. Stolz, Radiat. Protect. Dosim. **66**, 407 (1996).
18 U. Rieser, G. Hütt, M. R. Krbetschek and W. Stolz, Radiat. Meas. (1997, in press).
19 A. Lang and G. A. Wagner, Quaternary Geochronology (1996, in press).
20 M. J. Aitken and J. Xie, Quatern. Sci. Rev. **11**, 147 (1992).

Mineralogical Technologies in Archeology: Their Application to Grey-Minyan Ceramics of Troy VI

Görres M., Evangelakakis Ch., Kroll H.

Institut of Mineralogy, Westfälische Wilhelms-Universität, Münster,
Corrensstr. 24, D-48149 Münster, Germany
E-mail: Gorres@nwz.uni-muenster.de, Kroll@nwz.uni-muenster.de

Abstract. Modern X-ray powder diffraction as well as scanning and transmission electron microscopy techniques are used to characterize and compare ceramic firing products prepared in the laboratory with original bronze-age ceramics of Troy, Turkey.

1 Introduction

Ancient ceramics play an important role in archeological investigations as they carry information on former technological skills. 'High technologies' mark highly developed cultural activities, to the extent that some engineering techniques based on ancient technologies are still in use today, whereas others were lost in the past. Ancient urbanisation produced stratigraphic sequences of cultural activities which are marked by different types of ceramics. During an archeological excavation, large numbers of sherds may be found and classified according to their typical macroscopic profile (e.g. color, glance, porosity, special treatment etc.) in order to relate them to reference material. The archeologist can expand his classification profile with the assistance of the mineralogist in order to achieve a broader classification pattern. Mineralogical techniques related to materials science have produced large scale databases for solid state characterization.

In 1988, a new excavation campaign was started in Troy (Turkey) under the direction of Prof. Dr. M. Korfmann (Tübingen/Germany). The subject of our investigation is ceramics of Troy VI a-h (1700-1250 B.C.) containing a typical sort of grey colored ware, called Grey-Minyan Ceramics. The grey color implies that the ceramics were produced under reducing atmospheric conditions. Our experiments were therefore performed at Boudouards' equilibrium.

2 Experimental and analytical work

Clay material was collected from locations close to the excavation site. Part of it was fired at a local pottery (Ihson Aloco, Çanakkale) using traditional techniques. This material was compared to ancient ceramics by an archeologist (Dr. Frirdich, Tübingen) referring to their common macroscopic profiles. Those raw materials which, on firing, had produced a close match to the ceramics of Troy were used in extended experimental work performed in the laboratory based on traditional methods in processing ceramics (choice of raw material, processing up to styling of clay, furnace constructions) followed by local traditional potteries.

The goal of our experiments was to obtain information on the development of the mineral phases and the textures of the firing products as a function of temperature, and to compare these results with the findings made on Troy VI ceramics. The raw material was first characterized by chemical analysis, grain size distribution as well as X-ray powder diffractometry (XRPD). It was then processed under controlled laboratory conditions (removal of passive coarse grained minerals - ≥20μm - for better detection of newly developing phases, firing at Boudouard's equilibrium at 744°C, 840°C, 875°C, 940°C) to produce an internally consistent set of ceramic material for comparison and analysis. The change of the mineral content during firing (decomposition vs. production) was studied by XRPD and quantified by modern software technologies based on single crystal data (Rietveld Method). The mineral content of the raw material (illite, kaolinite, smectite, chlorite, and talc) becomes amorphous on firing whereas calcite decomposes. Quartz does not react up to 940°C. At 840°C, pyroxenes (diopside and enstatite) develop in increasing amounts. Olivine occurs as a minor component. Akermanite is stable between 744°C and 875°C only. Feldspars grow in the whole temperature range: sanidine is only a minor constituent, but anorthite grows in increasing amounts.

The newly grown minerals are extremly fine-grained (≈150nm). Therefore, they cannot be detected by polarized light microscopy. In addition, some or all of these mineral species may originally occur as coarse grained material in the matrix of ancient ceramics. Due to coincidences in the X-ray powder diffraction pattern, XRPD does not allow to distinguish between original and newly grown minerals.

In order to solve this analytical problem, the firing products as well as original ceramics of Troy VI were investigated by scanning (SEM) and transmission electron microscopy (TEM). The SEM images of the firing products display the alteration process of the clay minerals during heating. The clay minerals retain their platy habit up to ≈744°C. Between 744°C and 840°C melting sets in at the grain edges. As a result, the porosity of the sherd increases with increasing temperature. The melt areas appear glassy when viewed by SEM.

Using TEM, however, a large number of fine-grained crystals with grain sizes of only 50-250nm shows up and can be identified as a diagnostic mineral group (pyroxenes) from their electron diffraction patterns. A similar porous texture with pyroxenes grown in siliceous melt could be found in some well processed ancient ceramics of Troy VI. This finding points to a firing temperature of ≈840°C at reducing atmospheric conditions.

3 Conclusion

Mineralogical techniques such as scanning and transmission electron microscopy allow us to compare in detail the diagnostic textures and mineral contents of ancient ceramics with ceramics produced by firing clay material in the laboratory under controlled reducing or oxidizing conditions. Provided the starting materials are chemically and mineralogically similar, this comparison allows one to characterize the ancient firing procedures and to determine approximate temperatures which were achieved during firing.

Osteodensitometrical Studies and Elemental Analysis of Ancient Human Bones*

St. Jankuhn[1], H. Baumann[2], K. Bethge[2], T. Butz[1], R.-H. Flagmeyer[1], J. Hammerl[3], R. Protsch von Zieten[3], T. Reinert[1], I. Symietz[2], J. Vogt[1], and M. Wolf[3]

[1] Univ. Leipzig, Fak. f. Physik u. Geowiss., Abt. NFP, D-04103 Leipzig, Germany
[2] Univ. Frankfurt a. M., Inst. f. Kernphysik, D-60486 Frankfurt a. M., Germany
[3] Univ. Frankfurt a. M., Inst. d. Anthropologie, D-60323 Frankfurt a. M., Germany
*Supported by the BMBF under grant no. 03-BU9LEI.

Abstract. Quantitative digital radiography, ion beam methods, and X-ray fluorescence analysis are combined for the determination of bone mineral density and the analysis of major and trace elements of two Merowingian populations. Radial distributions of element concentrations are recorded to estimate *post-mortem* mineral exchange processes.

1 Introduction

In order to investigate age- and sex-related bone loss and metabolic bone disease up to now, several methods of osteodensitometry have been applied. To get more information about the relationship between biological development, health status, and living conditions of our ancestors, osteodensitometrical methods are coupled with ion beam techniques. First, the mortal remains are examined for their age and sex on the basis of morphological examinations. Then their bone mineral densities (BMD) are determined by quantitative digital radiography (QDR) and the elemental concentrations are measured using proton induced Gamma-ray emission (PIGE), proton induced X-ray emission (PIXE), Rutherford backscattering spectrometry (RBS) and X-ray fluorescence analysis (XRF). Among the skeletal regions which are indicators of fracture risk, the femoral neck including the so-called WARD's triangle is investigated in this study.

2 Samples

Several femurs from burial places from Bockenheim and Edesheim, Germany, were selected. After the excavation, the findings were separated from adherent soil and were pre-cleaned with water and brush. The specimens were dated across the 6th...8th centuries AD (Merowingian period) using the ^{14}C-method. For the analysis of trace elements the samples are prepared as follows: (i) Extraction of the neck zone using a diamond saw in order to pulverize it with a laboratory mill and to press the powder obtained into pellets or (ii) sawing off cross sections of about 1 mm thickness from the femoral shaft to record radial distributions of elements.

7 S. A. E. Johansson and J. L. Campbell, *PIXE: A Novel Technique for Elemental Analysis* (John Wiley & Sons, Chichester, 1988).
8 T. Reinert, Diploma Thesis (Univ. Leipzig, 1996).
9 J. R. Tesmer and M. Nastasi (eds.), *Handbook of Modern Ion Beam Materials Analysis* (MRS, Pittsburgh, PA, 1995).

Ruby Glass of the Baroque Period: An Interdisciplinary Study

I. Horn[2], D. von Kerssenbrock-Krosigk[1], G. Schulz[2], D. Syndram[1]

[1] Grünes Gewölbe, Staatliche Kunstsammlungen Dresden
[2] Institut für Anorganische und Analytische Chemie der Technischen Universität Berlin

1 The art historical approach

The making of gold-ruby glass vessels was developed around 1680 by Johann Kunckel in Potsdam. It had been previously known that the specific ruby-colour could be obtained by adding gold to the glass batch: Already in 1612, Antonio Neri published a gold-ruby recipe in Florence (Antonio Neri: "L'arte vetraria distinta in libri sette [...]", ch. 7,129). In France another Italian, Bernard Perrot, obtained the royal privilege to produce red glass in 1668. Since the production has its difficulties, the early Italian and French ruby glass consisted mainly of small, easy- to-handle products or applications on colourless glass vessels. Johann Kunckel, therefore, was the first to produce large, fully red coloured vessels. Since he also improved the mixture of the basic colourless glass, it was from then on possible to achieve brilliant, thick-walled vessels, the colour of which - if successful - gives the impression of being cut out of an oversized ruby gemstone. Kunckel himself reported his invention in a posthumous publication (Johann Kunckel von Löwenstern: "Collegium Physico-Chymicum Experimentale, Oder Laboratorium Chymicum [...]", 1st ed. by Johann Caspar Engelleder 1716, ch. 3,44). As to be expected, Kunckel's recipe did not remain secret for very long; it was passed on to other glassworks by disloyal assistants. There is ample written evidence of at least the temporary production of ruby glass in southern and northern Bohemia, in Bavaria as well as in Silesia. In Saxony, Johann Friedrich Böttger succeeded in about 1713 with his method of flashing a thin gold-ruby layer over colourless glass, a technique of which only a few examples survive in the Green Vault in Dresden. The still extant ruby glasses show that its production increased instantly after Kunckel's invention, arrived at its peak probably in the 1690's and slowly declined until the 1740's. It never became completely forgotten and several revivals occurred during the 19th century. The raised and cheapened production which now mainly took place in Bohemia deprived the gold ruby glass of some of its reputation - which nevertheless exceeds that of other glass colours up to modern times.

An overview of the history of red glass from antiquity until now shows its constant esteemed value. It can even be suggested - with every reservation against deterministic models - that since the earliest times glass manufacture has been working on a red, ruby-like tint which was not fully accomplished until the baroque gold ruby glass. The main part of the project deals with the baroque ruby glass itself. As many gold-ruby glass vessels as possible have been catalogued in order to be classified on the basis of their stylistic features. In comparison to one another, with colourless glass and other works of decorative art, the vessels have to be arranged chronologically; their common provenances have to be determined and their specific quality appreciated. The traditional art historical methods do not suffice to accomplish all these aims, in particular to determine the relation between groups with inferior stylistic properties. Yet these relations are highly important as they not only inform about the real number of glass works capable of ruby glass production, but also form the precondition for any further regional attributions. To solve these questions, the chemical composition has to be taken into consideration.

2 The chemical approach

The beginning of this undertaking was marked by the improvement of a quasi non-destructive sampling technique using diamond fine-grinding-files (DIAM method). Problems arose with the mass of the files being dependent on temperature and humidity; these factors have to be considered if - as is usually the

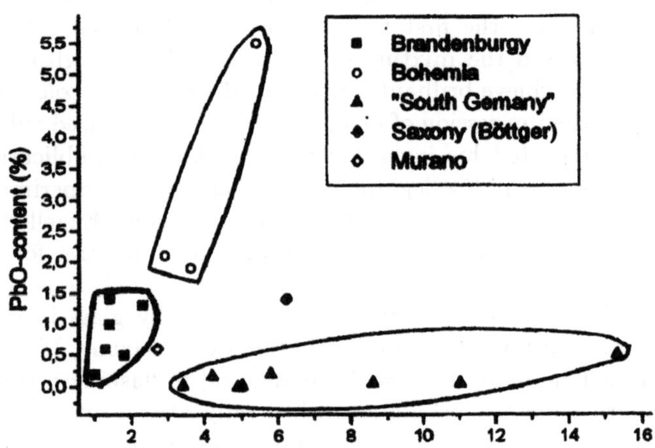

Fig. 1. Group formation; PbO-content versus CaO-content

case - several hours or days lie between weighing in the laboratory, sampling in the collection and the repetition of weighing in the laboratory. Cross-checking the mass-changes against some files which have been taken along but not used for sampling, an accuracy of about $\pm 5 \mu g$ (at a sample mass of 100 μg) can be achieved, which is sufficient for the archaeometrical purpose. The next step was to adapt the digestional and analytical procedures to the minute sample mass. Current procedures have been chosen (atomic absorption spectrometry, ion chromatography, voltammetry) to ease the approach and adaption. The results were verified by analyzing standard glasses and by inclusion of further methods (electron beam micro analysis). In a further step, the technique was applied on

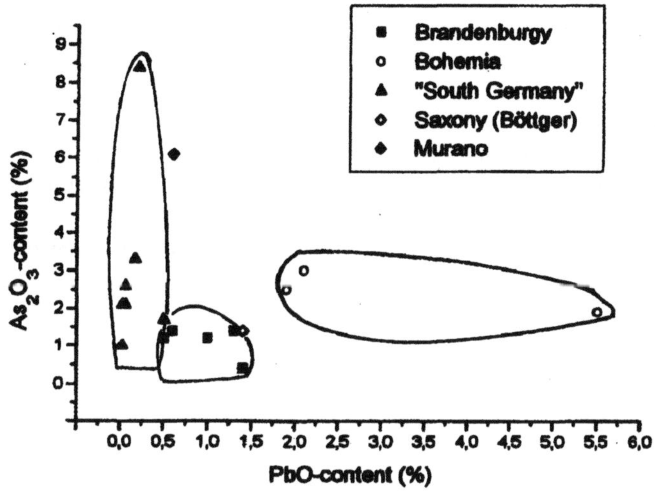

Fig. 2. Group formation; As_2O_3-content versus PbO-content

the baroque ruby glass vessels. To get an idea of the composition of glasses of different provenance (Brandenburg, Bohemia, Southern Germany), some representative vessels were chosen from the art historical part of the project, then sampled and analysed. In addition, written sources were evaluated for characteristic compositions. Although, naturally enough, the analytical results spread, it is possible to ascribe certain characteristics to each provenance (cf. figures 1 and 2). Due to the preciousness of the objects involved these data are unlikely to permit comprehensive statistics. Each sampling depends on the actual state of the object and must be obtained in cooperation between natural scientist and art historian.

3 The profit for the art historical evaluation

Surprisingly enough, baroque glass has never been subjected to a thorough chemical analysis comparable to that of antique, medieval and the distinct group of mainly Renaissance "facon de Venise" glass. The current project therefore has the character of a pilot scheme, on which the success of such investigations on the whole might be measured. Beyond just the analytical results, the project faces the task of an interdisciplinary teamwork between art historian and chemist, which is limited by the fact that art historical research seldom deals with fragments suitable to chemical investigation. Although the analyses for the current project therefore had to be reduced to a very small number, the collaboration is of significant use for the research of baroque gold ruby glass and can be summed up as follows:

- Similar to the stylistic patterns, chemical characteristics can be detected, which help to join or separate the various ruby glass groups.
- The still extant baroque ruby glass recipes can be included into the research providing information on certain temporal and regional developments of glass compositions.
- The relation to 'special cases' of ruby glass - the Saxonian flashed glass as well as the Italian and even Chinese ruby glass of the 18th century - can be defined.
- The analyses will be useful to identify imitations and forgeries.

Surface Change on Laser Irradiated Bioglass®

St. Szarska
Institute of Physics, Technical University of Wroclaw,
50-370 Wroclaw, Wyb. Wyspianskiego 27 , Poland

Abstract. The influence of 337 nm laser radiation and simulated physiological solution on the surface of the bioglass® is presented in this paper. As a result of these processes on the surface of the glass, a charged layer is formed. The applied measurement techniques were luminescence excitation spectra and exoelectron emission. The obtained results indicate the existence of the electron trap in the bioglass® structure. The nature of these centres is discussed in this paper.

1 Introduction

Bioglass® [1] is one of the materials which is mainly appropriated to reduce a disability and to prolong the human life. Its chemical composition is similar to the composition of bones. It belongs to the Na_2O-CaO-P_2O_5 system.

In an aqueous environment (physiological solution with pH = 7.4), the sodium ions are leached from the bioglass surface and replaced by H^+ ions, producing a silica-rich surface layer. First, an amorphous calcium phosphate layer is formed on the silica-gel layer due to the migration of ions from the bulk of glass. In 1-6 weeks the apatite agglomerate layers crystallize, which, in contact with the bone surface, join with collagen, forming the biological bonding. This process is called osteogenesis. As a result of the initial ion exchange process on the surface of the bioglass®, a negatively charged layer is formed [2]. The Ca,P - rich layer consists of hydroxyapatite of small crystallites and defective structures.

From the relevant literature it becomes obvious that the aqueous attack of glasses can be divided into two very general stages. The first is the selective leaching of mobile ions (e.g Na+, Ca2+) from the bulk of the solid through the surface and interface, where the pH remains below that necessary to cause the dissolution of the silica network. The second is a process of hydrolytic depolymerization in solutions containing a high concentration of OH ions (i.e pH> 9.5). However, a silicate glass exposed to an aqueous solution of pH<9 cannot be assumed to be free of local surface pH which can greatly exceed pH 9, except of course at the isoelectric point where the bulk solution pH and effective surface pH are equal. Thus, both of these reactions may occur simultaneously, and may be accompanied by ion exchange glass hydration or network hydrolysis [7].

Simultaneous action of the physiological solution and irradiation will affect the charge transfer kinetics. Before measurements, glass samples were irradiated by ultraviolet and visible laser radiation. Changes of the microstructure of the bioglass® surface resulting from irradiation and physiological solution treatment have been observed using the exoelectron emission and luminescence methods.

2 Experimental conditions

1. The luminescence was measured with the typical system consisting of a monochromator SPM-2, photomultiplier M12FC51, and recorder G1B1. The samples were excited with light of wavelength from the range 200 - 300 nm.

2. The OSEE current was registered by a secondary electron multiplier (10^{-18} A) in a vacuum chamber with 10^{-4} Pa. Measurements of optical-stimulated kinetics were carried out with an optical stimulation using 6 interference filters giving the wavelengths ranging from 225 to 325 nm. The parameters characterizing the OSEE decay curves were chosen as S_C surface charge: $S_c = \int_0^{\tau} I dt$. This method has been described elsewhere [3]. The number of quanta falling on the glass surface was constant for these measurements.

3. Glass samples were irradiated with an N_2 laser. The laser emitted light of wavelength 337 nm. The N_2 laser was working in pulse mode. Its frequency amounted to 30 Hz and time of one pulse was 10 μs. The power per surface unit in each pulse was about 10^5 W/cm^2*pulse.

4. The subject of investigations were two types of bioglasses produced by Jelenia Góra Optical Factory according to Hench [1] with the following composition (percentage by weight):

45 S 5 45 SiO_2, 24,5 Na_2O, 24,5 CaO, 6 P_2O_5

45S5.F 42,9 SiO_2,23,4Na_2O, 11,7 CaO, 16,3 CaF_2, 5,7P_2O_5

5. The samples were submitted to the action of the simulating physiological solution with pH equal to 7.4 for two weeks. Li and Zhang [4] reported that this ion-exchanging process results in a negatively charged surface. This high negative potential results in more Ca^{2+} ions in the electric double layer.

3 Result and discussion

The spectra of luminescence excitation are presented in fig.1. The spectrum of the luminescence excitation for pure 45S5 bioglass (fig.1 curve a) contains one band with a complex structure and a maximum at about 5.2eV (42* 10^3 cm^{-1}).

In comparison with the luminescence of the pure bioglass, the luminescence of the samples immersed in simulation physiological solution is more intensive but a maximum of the band of the luminescence excitation is the peak of Fig.1 curve b. The band with the maximum at 5.2 eV is related to oxygen vacancies. In the range of about 5.8 eV for the fluoride bioglass, the appearance of an additional maximum has been observed which has not been stated in the case of the 45S5 bioglass. Arbusov *et al.* [5] stated that this band may be associated with the decay of E' - centers.

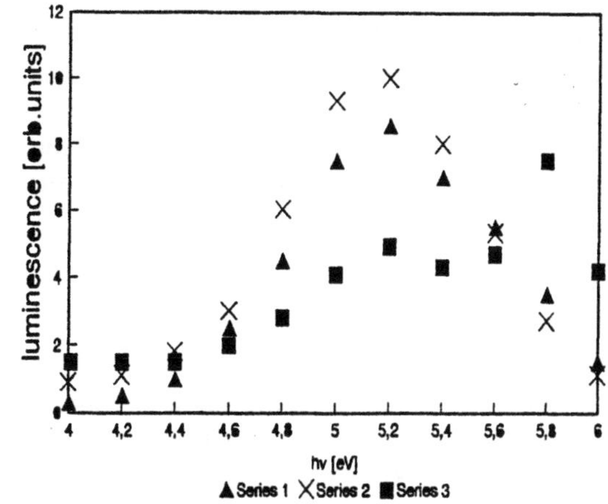

Fig.1 Luminescence excitation spectra for : 1-pure bioglass; 2 - 45S5 bioglass exposed in physiological solution; 3 - 45S5.F bioglass exposed in physiological solution (x4)

Line scanning of the surface of these glasses shows a pulsed fluctuation of exoemission current. In agreement with the physico-chemical conception of exoemission [6], such a fluctuation character results from the chemical interaction between bioglass and corrosion media (physiological solution), subsequent desorption of the products, appearance of the local plastic deformations and the formation of the microcracks. Local exothermic stress relief and evolution of the products resulting from chemical interaction lead to the emission of charges. Irradiation of bioglass produces a cation vacancy and the loss of an electron on an adjacent oxygen produces a V_1 center, defined as an O^- anion adjacent to a cation vacancy. The surface charge spectra calculated from optical stimulated exoelectron emission (OSEE) after exposure in simulated physiological solution for two weeks are presented in fig.2.

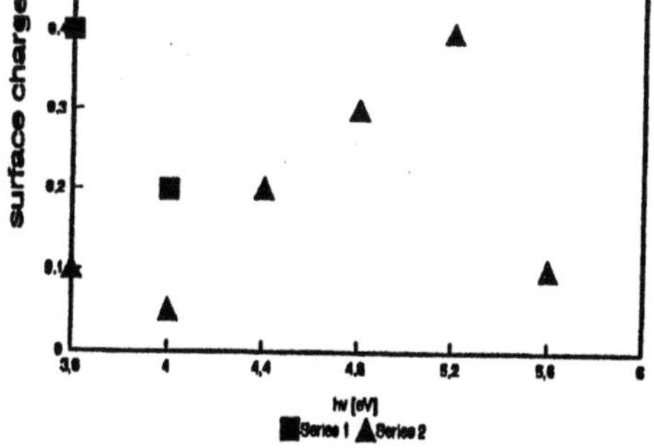

Fig.2 Surface charge of exoemission spectra (337 laser excitation) for: 1 - 45S5.F 2 - 45S5 bioglass. The samples were immersed in physiological solution for 2 weeks.

These samples were previously excited by a 337 nm laser. The maximum of the surface charge value of about 4.8 eV for the fluoride bioglass is probably connected with the absorption of water by fluoroapatite and formation of HPO_4^{-2} ions in the structure. For the 45S5 bioglass this maximum is shifted at about 5.2 eV. This result agrees with that for the luminescence spectra. Perhaps the source of emission is the same.

4 Conclusion

1. Exoemission activity of laser irradiated bioglass results from ionization of ensembles of bulk electronic defect centers formed due to the charge change in both of localized states of the conduction band and point to defects in the glass network like E_i^- centers, E' -centers.
2. Exoemission measurements indicate that the laser irradiation leads to filling of the electron traps present in the bioglass structure.

Acknowledgment

I would like to thank Dr. K. Marczuk for luminescence measurements.

References

1. Hench L.L., Splinter R.J., Allen W.C., Greenlee T.K., *J. Biomed. Mater. Res.* **2,** 1971, 117-141.
2. Kokubo T., Surface chemistry of bioactive glass ceramics, *J.Non-Crystalline Solids*, **120,** 1990, 138-151.
3. Szarska S. In: *Biomaterials & Human Body*, (eds. Ravaglioli A., Krajewski), 1992, pp.396-401, Elsevier P.C., Amsterdam.
4. Li P., Zhang F., *J. Non-Cryst. Sol.,* **119,** 1990, 112-117.
5. Arbusov V.I., Zatsepin A.F., Kortov V.S., Tolstoi M.N., Tyukov V.V., *J.Non-Cryst.Sol.* **134,** 1991, 208-217.
6. Krylova I.V., *Povierkhnost* **7,** 1993, 43-53.
7. Kruger A.A., In:*Surface and Near-Surface Chemistry of Oxide Materials* (eds. Nowotny J. and Dufour L.-C.), 1988, pp. 413-447. Elsevier S.P., Amsterdam.

Laser Cleaning

Excimer Laser Cleaning for Mediaeval Stained Glass Windows

Johanna Leissner
Fraunhofer-Institut für Silicatforschung, Bronnbach 28, D-97877 Wertheim
e-mail Leissner@isc.fhg.de, Germany

Abstract. The results of a pilot study show that KrF Excimer laser treatment seems to be a promising soft cleaning method for removing organic materials such as protective coatings, adhesives and biofilms from the surface of mediaeval stained glass windows.

1 Introduction

Mediaeval stained glass windows are suffering heavy deterioration due to increasing environmental pollution. Corrosion reactions take place and result in the formation of surface layers of gypsum and syngenite. Deposition of soot and dust, growth of biofilms and residual layers of organic polymers from previous restoration treatments cause additional problems for stained glass conservators. They have to use mechanical and chemical cleaning methods to remove these deposits to regain transparency and to decrease the corrosion progress. However, the cleaning methods used up to now include risk potentials of destroying the protective gel layer, e.g. by scratches that reach down to the core glass which again cause corrosion. Special problems are the removal of organic polymers and of biofilms.

In this pilot study, the aim was to prove the principle suitability of excimer laser radiation in the field of stained glass conservation.

Original historic glasses are unique pieces of art and their limited possibilities for quantitative analyses restrict their use as test specimens; therefore so-called standardized simulation glasses were used for the extensive test series required to examine the new cleaning technique.

2 Experimental procedure

For the experiments a KrF excimer laser (Lambda Physik, EMG 201 MSC) with a 248 nm wavelength was used. The energy flux varied between 1 and 4 J/cm^2, the number of pulses was varied in the range of 5 and 300. The pulse duration was 30 ns with a maximum repetition rate of 80 Hz.

Different simulation materials and one original piece from Erfurt Cathedral (13th century) were used for the laser treatments.
The variations of pretreatment, laser parameters and characterization methods are given in table 1.

Table 1. Parameters used for pilot study

type of pretreatment	laser parameter	characterization
*no treatment *glass with gel layer *glass with corrosion crust *corroded glasses with a) ORMOCER coating b) Paraloid B 72 coating c) SZA paint consolidant *mediaeval glass (13 c.)	energy density $F= 1\text{-}4\ J/cm^2$ pulse number J=5-300 pulses area of irradiation $A \cong 1 mm^2$	FT-IR microscope (reflection, transmission) SEM profilometer light microscope

3 Results and discussion

The most promising results were obtained for the removal of organic materials from the glass surface: a defined and homogeneous ablation of these layers could be initiated using very low energy densities ($\leq 1\ J/cm^2$) without any damaging interaction of the laser radiation with the protective gel layer or the core glass. IR (infrared) spectra can be used to control the ablation process. A pure organic coating of Paraloid B 72 (copolymer of methyl acrylate and ethyl methacrylate) demanded the lowest pulse numbers to start the ablation process. For the removal of ORMOCER coatings (inorganic-organic hybrid composite materials), pulse numbers higher than 10 are required. The transmission IR spectra clearly show (Figure 1) the disappearance of the CH absorption bands between 2900 and 3100 cm^{-1} (aromatic and aliphatic hydrocarbon groups from the ORMOCER lacquer) when pulse numbers exceed 10. The reflection IR spectra (Figure 2, curve A) prove that for pulse numbers less than 10, the ORMOCER coating is not detached. Increasing pulse numbers result in the complete ablation of the coating, but the protective gel layer remains intact (appearance of Si-O-Si band at 1100 cm^{-1}, Figure 2, curve B). Further increase in the energy densities and pulse numbers causes the ablation of the gel layer. The typical absorption bands at 1050 and 930 cm^{-1} are then visible. However, their intensities are low because of higher surface roughness caused by the laser beam (loss of signal intensity due to scattering, Figure 2, curves C,D).
The removal of different types of corrosion crusts was possible. However, the ablation was not homogeneous and precise since the corrosion crust is inhomogeneous itself. The energy threshold required to start the removal of the inorganic corrosion crust is in the same range as for the removal of bulk

glass and protective gel layer. The probability of unintentional ablation of the gel layer and bulk glass is very high.

Tests on small parts of an original 13th century piece of a stained glass window from Erfurt Cathedral (Germany) showed that the results obtained from simulation glasses and on originals are nearly congruent: above 2.5 J/cm^2 removal of corrosion crusts was possible and thus the original piece regained transparency. Higher energies caused remelting of the glass surface which must be avoided.

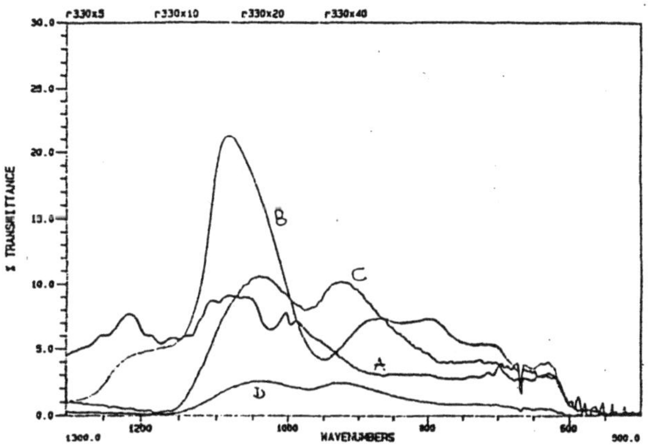

Fig. 1. IR transmission spectra of an ORMOCER coated glass after laser treatment with different pulse numbers (A: 5, B: 10, C: 20, D: 40)

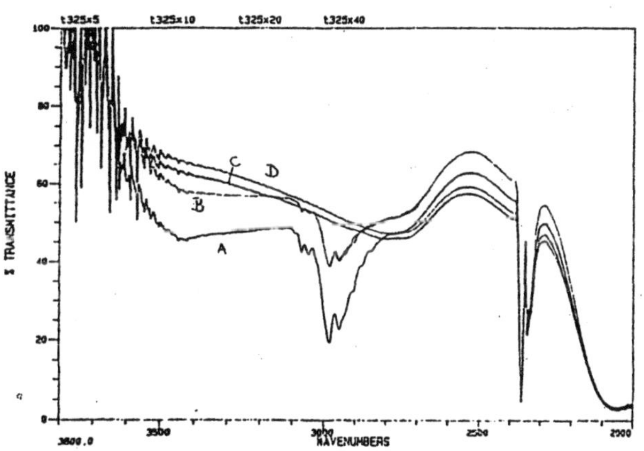

Fig. 2. IR reflection spectra of an ORMOCER coated glass after laser treatment with different pulse numbers (A: 5, B: 10, C: 20, D: 40)

4 Conclusions

The KrF excimer laser treatments have potential as a soft cleaning method for the removal of organic layers which are difficult to remove with chemical or mechanical tools. These results set up new perspectives in the conservation and preservation practice: New polymeric materials offering better protection against deterioration could be applied without risk of irreversibility which is a main concern of conservators and restorers.

More detailed studies are, however, necessary to provide a data base on the long term effects of UV Laser irradiation before the new technology can be recommended to the workshops.

References

1 H. Römich and D.R. Fuchs; BMFT Verbundprojekt BAU 5026 C/4 (1992).
2 J. Leissner, W. Barkhausen, K. Wissenbach and D. R. Fuchs, BMFT Verbundprojekt, BAU 5026 C/4 (1994).
3 H. Römich and D. R. Fuchs, Bol. Soc. Esp. Ceram. Vid. 31-C 137-141 (1992).

Laser Cleaning with a Nd:YAG Laser

Paraskevi Pouli, David C. Emmony

Department of Physics, Loughborough University, Leics, LE11 3TU, UK
E-mail: P.Pouli@lboro.ac.uk, D.C.Emmony@lboro.ac.uk

Abstract. In this paper the practical aspects of laser cleaning a small area of the exterior of Lincoln Cathedral will be presented. Surface analysis results and the optimum cleaning rates will be discussed with the objective of the preservation of the limestone patina.

1 Introduction

Laser technology has been used during recent years for the treatment and restoration of our historic and artistic heritage. Much research has been carried out to understand and explain the interaction of laser radiation with these materials.[1] Nd:YAG laser radiation (at 1.064 μm) has proved to be an efficient "tool" for the removal of the polluted black encrustation on stonework.

2 Weathering of Stone

Stonework changes inevitably when it is exposed to the environment. The passage of time and the weather can cause alterations in the structure and significant loss of material or detail. Our major concern is the black encrustation which forms on stone masonry.

2.1 The Black Crust and the Patination Layer

The surface layer on limestone is a complicated feature found mainly in heavily polluting environments. It appears as a black layer that covers extended areas especially those well protected from direct rain washing. Surface analysis has shown that it consists of gypsum crystals ($CaSO_4$ $2H_2O$) with carbonaceous particles from fuel combustion. The main mechanism responsible for this encrustation is a combination of dry and wet deposition of all the combustion products discharged into the atmosphere on the fragile and accommodating limestone and replacement of the calcite in the limestone by gypsum, through the sulphation mechanism from the interactions of stone with rain-water and atmospheric pollutants.

Immediately after removal of the black crust by a laser a yellowish surface layer is visible that researchers have agreed be called the "patina". Surface analysis (XRD, SEM) has shown that the patina consists mainly of gypsum ($CaSO_4$ $2H_2O$) and lime (CaO). This supports the idea that this layer is a physical reaction of the stone to its natural environment. Conservators agree that the patina is the best laser cleaned surface [2], mainly because it protects the underlying stone from further weathering.

2.2 The Laser Cleaning

The laser cleaning process is self-limiting because it depends upon the difference in the absorption, at the Nd:YAG wavelength, between the dark encrustation and the light underlying limestone. Therefore the laser-solid interaction ceases as soon as the black layer has been removed.

3 The Lincoln Project

During the winter of 1995-1996 a project has been carried out at Lincoln Cathedral to clean part of the exterior of the building. The aim was to adjust the parameters of the laser cleaning system to outdoor conditions and optimise the cleaning efficiency of large areas.

3.1 Pollution Problems

The Cathedral is a 12th century building made of local oolitic limestone. This type of stone is highly porous and permeable. It is extremely delicate and easily soluble in rainwater. As a result it is very susceptible to all the deteriorating mechanisms. Moreover, the Cathedral is built on top of a hill in a rural region surrounded by several coal-fired power stations which have contributed to its weathering, providing a polluting environment to the building particularly over the last century.

Apart from the contamination of Lincoln masonry with black crust which results in severe alteration of the morphology of the stonework and the aesthetic appearence of the Cathedral, other weathering phenomena have occured such as [3]:

Blistering, in the form of volcano-like structures caused by gypsum crystal growth on the surface of the stone. The severity of this kind of damage is dependant on the type and orientation of the stone. Characteristic is the loosening and forcing apart of the surface layers revealing a fresh surface for further errosion.

Rain washing, in the form of whitish areas in regions directly exposed to rain, where it is observed that severe thinning and dissolution of the limestone takes place.

Cracking, due to frost action, caused by abrupt freeze-thaw cycles

3.2 On Site

A temporary enclosure was constructed around the doorway of the south transept cript of the Cathedral. This region is severely corroded and covered with a thick layer of black crust, with representative examples of the major deteriorating problems.Access to the cleaning site was quite controlled and thus health and safety rules could be easily followed. The laser system with beam manipulator arm, power supply and dust extractor were all sited in the enclosure.

The laser, a Spectron SL450 Pulsed Nd:YAG system, with maximum output pulse energy of 1 J at 10 Hz, had to be maintained at >10 °C, by placing a small portable heater inside the enclosure, while an artificial daylight lamp solved the problem of insufficient light. The cleaning pulse energy was up to 300 mJ at 7 Hz.

3.3 The Results

As we can see in the photographs the Nd:YAG laser is a very gentle and sensitive tool for cleaning the fragile Lincoln stonework. It proved to be capable of removing the black dirt layer while leaving the patination layer unaffected and revealing carving details. Delicate areas, like well formed blisters and exfoliation, were treated very carefully to prevent further material loss.

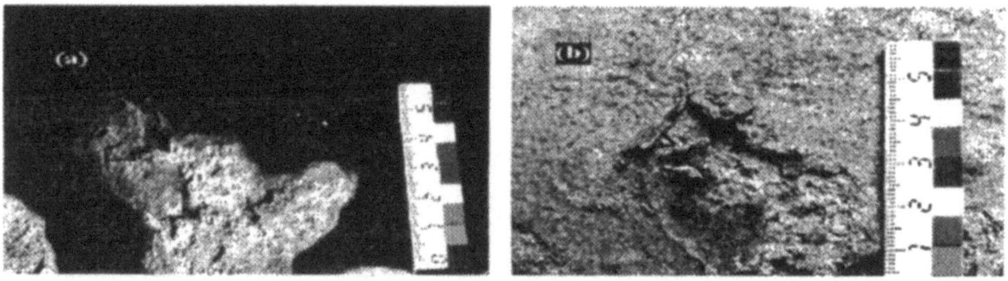

Fig. 1. A well formed blister on lincoln masonry before (a) and after (b) laser cleaning

4 Conclusions

The laser cleaning process has distinct advantages over conventional methods as:
(a) it is very convenient and clean and it can be maintained in a simple sheltering enclosure without any important difficulties.
(b) cleaning could continue in poor weather conditions (snow) and in very low temperatures (several degrees below freezing) without any associated dangers due to frost action.
(c) it is simple and safe to use when straightforward precautions (limited access and protection goggles) are employed.

The laser is the optimum tool for fine, delicate, safe and practical cleaning.

Acknowledgements

The authors would like to thank K. Beadman, M. O'Connor and J. Scarrow of Lincoln Cathedral Works Department and M. Cooper and J. Larson of National Museums and Galleries on Merseyside, Liverpool.

References

1 M.I. Cooper, D.C. Emmony, J.H. Larson, "*Characterisation of laser cleaning of limestone*", Optics and Laser Technology, 27 (1994).
2 V. Vergès-Belmin, C. Pichot, G. Orial, "*Elimination de croûtes noires sur marbre et craie: à quel niveau arrêter le nettoyage?*", Proceedings of the International Congress on the Conservation of Stone and other materials, (Paris, 1993).
3 R.N. Butlin et al, "*A study of the degradation of building materials at Lincoln Cathedral, Lincoln, England*", Proceedings of the 6th International Congress on Deterioration and Conservation of Stone, (Torun, 1988).

Pattern Recognition, Including Digital Evaluation

Optical Correlation Methods in Epigraphy

Nazif Demoli
Institute of Physics, PO Box 304, 10000 Zagreb, Croatia
E-mail: demoli@bobi.ifs.hr

Abstract. An overview of the recognition problem and the possible approaches for solution in epigraphy is given. Examples on the averaging of patterns, feature extraction, and the system configuration design are given to demonstrate the main issues.

1 Introduction

Character or sign reading has been the matter of concern in the field of optical pattern recognition, firstly to demonstrate the technique [1], and later to study inscriptions. The identification and classification procedures as well as the feature extraction can be successfully accomplished by using coherent optical processor systems and correlation experiments. The correlation measurements in epigraphy are associated with especially challenging problems due to a large number of the distorted samples of an object of interest. These distortions influence the design of rather sophisticated correlation filters (CFs) that are capable of decreasing the sensitivity between different samples of an object (true-class or in-class objects), while keeping the discrimination of other objects (false-class or out-of-class objects) on a reasonably high level.

Optical correlation methods are based on the Wiener-Khintchine theorem,

$$\mathrm{s}(x,y) \bullet \mathrm{h}^*(x,y) = F^{-1}\left\{\mathrm{S}(u,v) \cdot \mathrm{H}^*(u,v)\right\}, \qquad (1.1)$$

where $\mathrm{S}(u,v)$ and $\mathrm{H}(u,v)$ represent the Fourier transform (FT) of the input and reference signal, respectively, $\bullet$ denotes the correlation operation, F denotes the FT operation (-1 denoting 'inverse'), and an asterisk denotes a complex conjugate. The coordinates of the input and correlation planes are denoted with (x,y), while the coordinates of the Fourier space are denoted with (u,v). According to the theorem, the correlation of two complex signals is equal to the inverse FT of the product of their Fourier spectra. As seen from the right-hand side of equation (1.1), one of the two spectra must be complex conjugated.

Optical implementation of the recognition scheme described by the relation (1.1) is based on the FT property of a spherical lens. It is well known [2] that a spherical lens is performing the FT operation from one focal plane to the other, when illuminated with the collimated coherent laser light. The typical so called 4f coherent optical system is shown in Fig. 1. Three planes (input P_1, frequency P_2, and output P_3) and two lenses (FTL_1 and FTL_2) allow several functions to be achieved at the output of the system: (i) imaging an input signal with the possibility of spatial filtering in the frequency domain; (ii) correlation of the input and reference signals by means of optical multiplication in the frequency domain: the classical frequency plane correlator system; and (iii) correlation of the input and reference signals by means of the optical addition and detection in the frequency domain: the joint transform correlator system.

The output plane of an optical correlator consists of peaks of light. The presence of a peak of light indicates the presence of the reference signal in the input, the location of the peak of light denotes the location of the reference signal in the input, and the peak intensity shows the similarity degree between the reference and input signals.

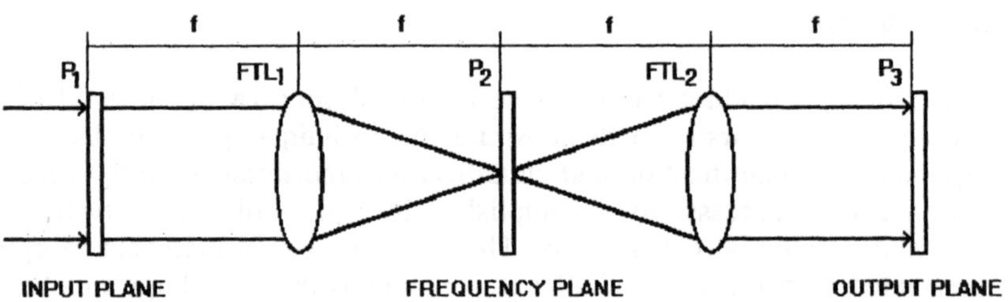

1:	h	→	F	→	$H(\cdot F)$	→	F	→	$h(\otimes f)$
2:	s	→	F	→	$S \cdot H^*$	→	F	→	$s \bullet h^*$
3:	s + h	→	F	→	$\lvert S+H \rvert^2$	→	F	→	$s \bullet h^*$

Fig. 1. The classical 4f coherent optical system.

Epigraphy is defined as a study that deals with deciphering, interpreting, and classifying inscriptions, especially ancient inscriptions.

2 Definition of the Problem

Inscriptions play one of the most important roles on the cultural-historical stage of the human civilization. Fig. 2 shows a hypothetical volume, called inscription space-time spreading volume, where the letters A, B, C, etc. denote the areas influenced by different inscriptions. A combination of the space-development of inscriptions with the time-slices is shown in a picturesque way. The general idea is to cover all possibilities due to the development or other changes by characterizing the spots of Fig. 2.

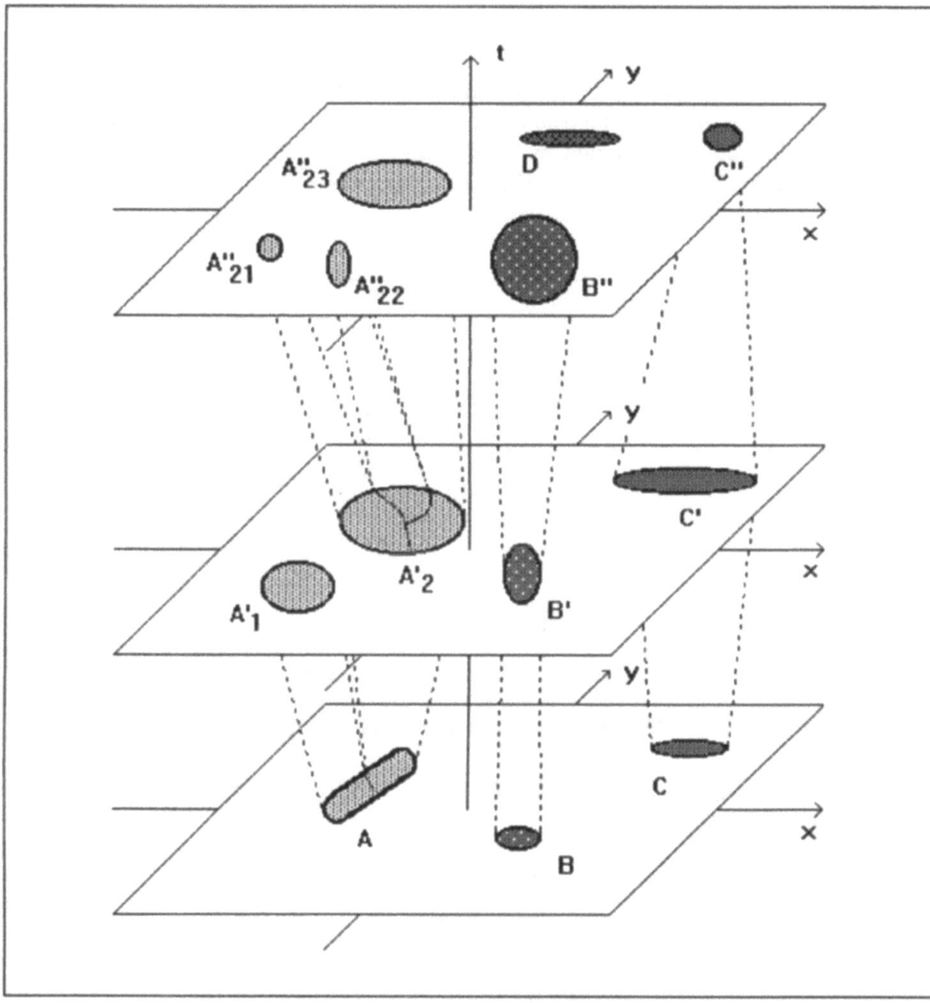

Fig. 2. Inscription space-time spreading volume.

Suppose we have an unknown written document. To learn about the time and place of writing, about the writing style and writing tool, and to decipher the writing contents, we must locate the document in the space-time volume shown in Fig. 2. To realize that accurately, the spots visible in Fig. 2 must be characterized. These spots represent different classes. The first step is the definition of the classes. Since classes can have many samples, we choose the best representatives of classes to cover the possible variations. Then the common v. different features are extracted in the selected feature space and the statistical properties are calculated. On the basis of the correlation measurements, a probability density cloud within the space-time volume of Fig. 2 is then defined containing the location probability of the unknown document.

3 Notation

Let the training set consists of N images and that each image contains $d=d_1 d_2$ pixels. For the *i*th training image of the true-class, denoted by $x_i(m,n)$, we first form its two-dimensional discrete FT, denoted by $X_i(k,l)$, and then by lexicographic reordering we form a vector, see Fig. 3. In Fig. 3, *F* denotes the FT operation and the superscript *T* denotes the transpose operation.

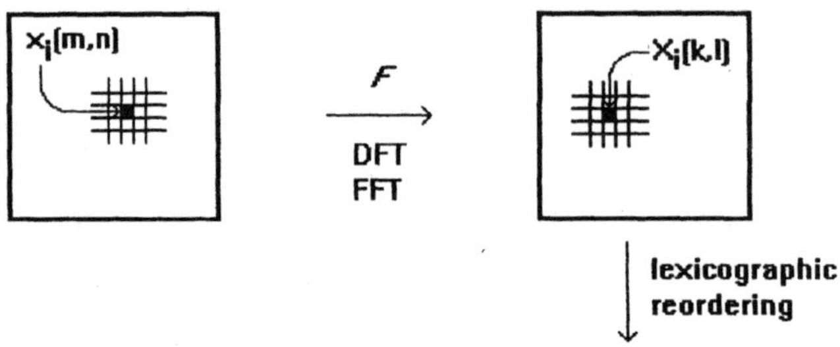

$$[\, X_i(1,1) \quad X_i(1,2) \quad \ldots \quad X_i(k,l) \quad \ldots \quad X_i(d_1,d_2)\,]^T \equiv \mathbf{x}_i$$

Fig. 3. Forming a true-class vector.

The *i*th training vector, $\mathbf{x}_i = [\, X_i(1) \quad X_i(2) \quad \ldots \quad X_i(d)\,]^T$, can be also represented as a diagonal matrix:

$$\mathbf{X}_i = \begin{bmatrix} X_i(1) & 0 & \ldots & 0 \\ 0 & X_i(2) & \ldots & 0 \\ \ldots & \ldots & \ldots & \ldots \\ 0 & 0 & \ldots & X_i(d) \end{bmatrix}, \tag{3.1}$$

Thus, the training-set true-class matrix **X** with N vectors, $\mathbf{X} = [\, \mathbf{x}_1 \ \mathbf{x}_2 \ \ldots \ \mathbf{x}_N\,]$, contains the necessary information about true-class samples in d rows and N columns:

$$\mathbf{X} = \begin{bmatrix} X_1(1) & X_2(1) & \ldots & X_N(1) \\ X_1(2) & X_2(2) & \ldots & X_N(2) \\ \ldots & \ldots & \ldots & \ldots \\ X_1(d) & X_2(d) & \ldots & X_N(d) \end{bmatrix}. \tag{3.2}$$

The training-set false-class matrix **Y** is obtained following the same procedure as for **X**: $y_i(m,n) \rightarrow Y_i(k,l) \rightarrow \mathbf{y}_i$ or $\mathbf{Y}_i \rightarrow \mathbf{Y}$.

The noise spectral density matrix $\mathbf{P}_n$,

$$\mathbf{P}_n = \begin{bmatrix} P_n(1) & 0 & \dots & 0 \\ 0 & P_n(2) & \dots & 0 \\ \dots & \dots & \dots & \dots \\ 0 & 0 & \dots & P_n(d) \end{bmatrix}, \tag{3.3}$$

is a diagonal matrix, where $P_n(j)$, $j \ni [1, 2,\dots d]$, is power spectral density of the additive stationary input noise.

The average values are given in following:

(a) Average true-class vector

$$\mathbf{m}_x = \frac{1}{N}\sum_{i=1}^{N} \mathbf{x}_i , \tag{3.4}$$

(b) Average true-class matrix

$$\mathbf{M}_x = \frac{1}{N}\sum_{i=1}^{N} \mathbf{X}_i , \tag{3.5}$$

(c) Average true-class power spectrum matrix

$$\mathbf{D}_x = \frac{1}{N}\sum_{i=1}^{N} \mathbf{X}_i^* \mathbf{X}_i . \tag{3.6}$$

4 Correlation Measures

If we multiply the transposed and complex conjugated matrix **X** with the filter vector **h**, we obtain a vector **u**:

$$\mathbf{X}^+ \mathbf{h} = \mathbf{u} , \tag{4.1}$$

containing the desired training-set correlation peak values. The superscript + denotes the conjugate transpose. Relation (4.1) represents a set of N linear equation with d unknowns. Since d >> N, many solutions for creating the filter vector **h** exist that satisfy the conditions of Eq.(4.1). For example, **h** was assumed to be a linear combination of the training-set vectors,

$$\mathbf{h} = \sum_{i=1}^{N} a_i \mathbf{x}_i , \tag{4.2}$$

known as the conventional synthetic discriminant function (SDF) [3], where the coefficients a_i, $i = 1, 2, \dots, N$ have to be determined to satisfy Eq. (4.1).

Several correlation measures, such as:

(a) Average correlation energy

$$ACE = \mathbf{h}^+ \mathbf{D}_x \mathbf{h}, \tag{4.3}$$

(b) Average correlation peak energy

$$ACPE = \mathbf{h}^+ \mathbf{m}_x \mathbf{m}_x^+ \mathbf{h} = |\mathbf{h}^+ \mathbf{m}_x|^2, \tag{4.4}$$

(c) Output noise variance

$$ONV = \mathbf{h}^+ \mathbf{P}_n \mathbf{h}, \tag{4.5}$$

(d) Average similarity measure

$$ASM = \mathbf{h}^+ \mathbf{S}_x \mathbf{h}, \tag{4.6}$$

where $\mathbf{S}_x = \frac{1}{N}\sum_{i=1}^{N} (\mathbf{X}_i - \mathbf{M}_x)^* (\mathbf{X}_i - \mathbf{M}_x)$, are often used measures to describe CFs in optical pattern recognition applications.

The meaning of the ACE measure, for example, is shown in Fig. 4, where from the normalized energy distributions for the reference, true-class, and false-class signals, it is evident that the average correlation energies for the true-class and false-class objects must be related as: $ACE_x >> ACE_y$.

Average Correlation Energy

$$ACE_x = \sum_{j=1}^{d} |M_x(j)|^2 |h(j)|^2 \quad \text{True-class}$$

$$ACE_y = \sum_{j=1}^{d} |M_y(j)|^2 |h(j)|^2 \quad \text{False-class}$$

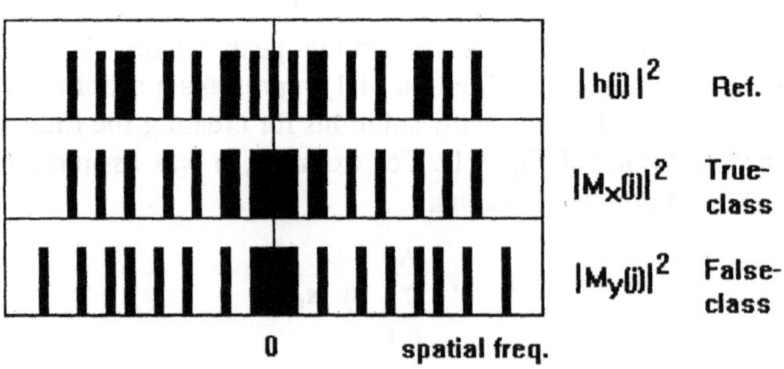

Fig. 4. An example given for the ACE measure.

5 Correlation Filter Design

Designing of the CF is one of the crucial tasks in correlating two signals when the frequency plane correlator of Fig. 1 is used. Ideal filter ought to recognize any distorted version of the reference object and reject all other objects. Since the ideal filter does not exist, the best we can do is to realize an optimum CF. The CF performance optimization is usually a result of the trade-offs between various correlation measures [4,5]. For example, by maximizing the criterion ACPE / (ASM_x + ACE_y), the maximum average correlation height (MACH) filter is obtained [6],

$$\mathbf{h} \propto (\mathbf{S}_x + \mathbf{D}_y)^{-1} \, \mathbf{m}_x , \tag{5.1}$$

with sharp correlation peaks for the true-class objects, and with the simple class statistics and the simple matrix inversions involved.

Regarding types, CFs were classified [7] into a basic set consisting of three main group of CFs: classical matched filters, phase-only filters, and inverse matched filters, due to the amplitude and phase information contents. Generally, the phase part of the filter should be either matched or binary matched to the phase part of the reference spectrum. On the other hand, the amplitude part of the filter can be arbitrarily designed. According to the analysis involving the feasibility estimate measure, a CF called the quasi phase-only filter (QPOF) [8], belonging to the group of phase-only filters, is found to be very effective in optical pattern recognition [7].

6. Application Examples

6.1 Averaging of Patterns

Consider an input consisting of two characters of the same class, such as shown in Fig. 5(a), i.e.

$$h(m,n) = \sum_{i=1}^{2} h_i(m,n) \otimes \delta(m-m_i, n-n_i) \; . \tag{6.1}$$

Then, its FT is equal

$$H(k,l) = \sum_{i=1}^{2} H_i(k,l) \cdot e^{-i2\pi(m_i k + n_i l)} \; , \tag{6.2}$$

and the power spectrum:

$$|H(k,l)|^2 = \sum_{i=1}^{2} \sum_{j=1}^{2} H_i(k,l) H_j^*(k,l) \cdot e^{-i2\pi[(m_i-m_j)k+(n_i-n_j)l]} \; . \tag{6.3}$$

The goal is to form the average pattern $h_{av}(m,n)$, shown in Fig. 5(b), where the delta functions and complex exponential functions, which influence the modulation in the power spectrum described by the Eq. (6.3), are not desirable. A solution for the optical averaging was proposed as a three-step procedure and demonstrated on the classification of the ancient handwritten Hebraic characters [9]. Applying that procedure on the input shown in Fig. 5(a), first a reference point sources must be added to the each character at exactly the same relative positions, as shown in Fig. 5(c). Then, the FT of the new input must be holographically recorded. In the final step, the average pattern is obtained at the output of the optical system of Fig. 1 with the hologram as input and the prefiltering performed in the plane P_2. An improved version of the coherent optical averaging technique was reported recently [10].

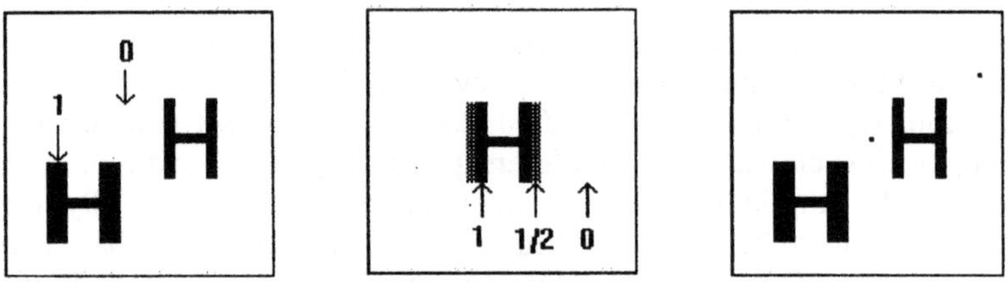

(a) (b) (c)

Fig. 5. Averaging of two characters: (a) input, (b) average character, (c) new input.

6.2 Feature Extraction

To demonstrate the feature extraction, the Chinese character 'rice' is shown in Fig. 6(a) with resolution 32x32. Chinese characters are large in number (more then 5400) and of complex shapes. To solve the input problem of Chinese characters, the optical recognition system was developed using accumulated stroke features [11]. For the character 'rice', the feature matrix is shown in Fig. 6(b), where 1 means 'no stroke', 2: 'vertical stroke', 4: 'stroke along 135^0 direction', 8: 'horizontal stroke', and 16: 'stroke along 45^0 direction'.

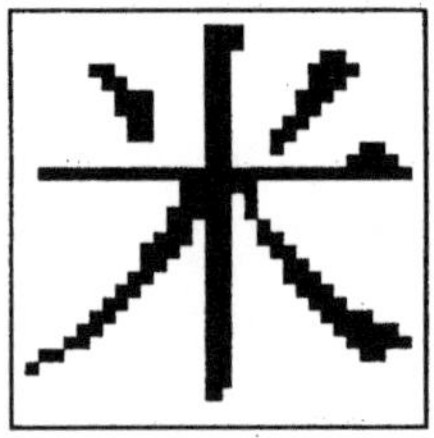

1	1	1	2	2	1	2	1
1	4	4	2	2	16	16	1
1	1	2	2	2	16	8	8
8	8	8	8	8	8	8	8
1	1	16	2	2	4	1	1
1	16	16	2	2	4	4	1
16	16	1	2	2	1	4	4
1	1	1	2	2	1	1	1

(a) (b)

Fig. 6. The 'rice' character: (a) the image and (b) the feature matrix.

6.3 System Configuration

Cuneiform inscription (CI) signs have been recently analyzed using holographic techniques. The investigation was concerned with two main problems: (i) a high-resolution archival storage [12] and (ii) a characterization of the CI signs [13]. For the characterization of the CI signs a multifunctional optoelectronic system has been developed. A functional scheme of the system is given in Fig. 7, where **h** denotes a reference signal, **s** an arbitrary input signal, and capital letters denote the FT of spatial functions. Generally, the optical system, supported by the digital image processing (DIP) systems and CCD cameras, has been advanced for the performance of various procedures such as the detection, Fourier analysis, feature extraction, coherent optical averaging, and correlation experiments. The basic 4f configuration of the system was extended to 4f+(5f)+(5f), thus enabling the in-line optical prefiltering. The spectrum modification **H'** at plane P_2 can be described by the convolution of the FT of the input signal with the FT of the input aperture, multiplied with the frequency plane spatial filter placed at plane P'_1 , and then further convolved with the image plane spatial filter placed at plane P'_2.

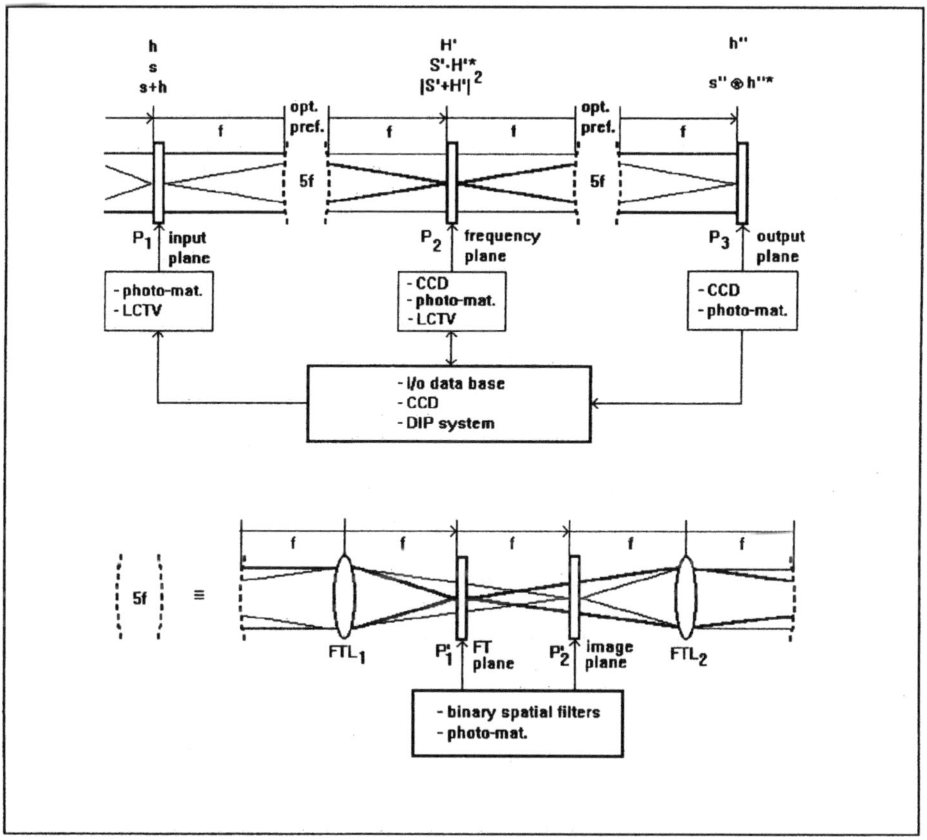

Fig. 7. Functional scheme of an experimental device.

7 Conclusions

Optical correlation methods represent a powerful tool for various two-dimensional recognition applications. However, to use that tool efficiently, the concrete problems such as the synthesis and the implementation of the optimum correlation filters must be solved. These problems are especially challenging in epigraphy due to a large number of objects of interest as well as a large number of object distortions. For example, the total collection of the cuneiform inscription tablets comprises approximately half a million pieces spread all over the world. The creation of a modest and efficient knowledge data-base for characterizing the cuneiform inscriptions is a huge task. The best way of realizing such tasks is the utilization of the optical methods in conjunction with the diverse alternative approaches.

Acknowledgments

The work reported here was partially supported by the Ministry of Science and Technology of the Republic of Croatia, and partially by the Ministry of Education, Research, Science, and Technology of the Federal Republic of Germany.

References

1 A. Vander Lugt, IEEE Trans. Inform. Theory **IT-10**, 139 (1964).
2 J. W. Goodman, *Introduction to Fourier Optics* (McGraw-Hill, New York, 1968).
3 C. F. Hester and D. Casasent, Appl. Opt. **19**, 1758 (1980).
4 B. V. K. Vijaya Kumar and L. Hassebrook, Appl. Opt. **29**, 2997 (1990).
5 V. Laude and Ph. Refregier, Appl. Opt. **33**, 4465 (1994).
6 A. Mahalanobis, B. V. K. Vijaya Kumar, S. Song, S. R. F. Sims, and J. F. Epperson, Appl. Opt. **33**, 3751 (1994).
7 N. Demoli, Optics & Laser Technology **26**, 119 (1994) & **28**, 215 (1996).
8 N. Demoli, Appl. Opt. **26**, 2058 (1987).
9 J. M. Fournier, in *Applications of Holography and Optical Data Processing*, (E. Marom et al., Eds., pp. 533-540, Pergamon Press, London, 1977).
10 N. Demoli, U. Dahms, H. Gruber, G. Wernicke, Optik **102**, 125 (1996).
11 B.-S. Jeng, Opt. Eng. **28**, 793 (1989).
12 G. von Bally, D. Vukicevic, N. Demoli, H. Bjelkhagen, G. Wernicke, U. Dahms, H. Gruber, W. Sommerfeld, Naturwissenschaften **81**, 563 (1994).
13 N. Demoli, H. Gruber, U. Dahms, G. Wernicke, Appl. Opt. (in press).

Cuneiform Recognition Experiments: Coherent Optical Methods and Results

Günther Wernicke, Nazif Demoli*, Hartmut Gruber, Uwe Dahms
Humboldt-Universität zu Berlin, Institut für Physik, Invalidenstr. 110, 10115 Berlin
*Institute of Physics at the University of Zagreb, 4100 Zagreb, Croatia

Abstract. The characterization of cuneiform signs by means of detection, feature enhancement and extraction, and identification is of special interest. To analyse cuneiform inscriptions, a multifunctional extended optical correlator and a joint transform correllator have been advanced. The use of methods minimizing the in-class sensitivity while maximizing the out-of-class discrimination ability is described.

1 Introduction

Clay tablets with cuneiform inscriptions are ancient documents. These handwritten inscriptions are three-dimensional information carriers. The font of cuneiform signs consists of up to 600 elements depending on the time and location. There is a large number of objects with a high cultural historic value. Some of these tablets are partially destroyed and different parts are sometimes stored at different places. So it follows that it is necessary to store these objects and their information content with the smallest possible loss of information, to reconstruct objects from fragments, and to identify inscriptions objectively depending on specific charcteristics, e.g. time, place, writer, or writing school.
For the solution of these problems it is possible to use methods of coherent optics. Coherent optics allow the possibility to use optical and digital techniques for data storage, data filtering, and pattern recognition. The goals of our work were to construct and test holographic / coherent optical setups for the determination of correlations of cuneiform signs, and to identify cuneiform signs written by the same writer. In relation to these goals it was necessary to optimize the coherent-optical techniques depending on the objects of interest.

2 Correlation Experiments

The background of the experiment is the determination of the similarity of input information with a reference. The measured value is a correlation peak, its intensity, peak height, and its half width value. These are influenced by the type of correlator or the input information. In our investigations we used an extended optical correlator of the Van der Lugt type in connection with a digital image processing system [1], [2], [3], and a joint transform correlator with liquid crystal displays [4], [5]. The

input information, taken from the cuneiform tablet, was a grey value distribution, a binarized distribution, and digitally or optically preprocessed distributions.
The input objects of different kinds are compared with reference objects. The following objects of interest were investigated:

Input	**Reference**
Single wedge	Single wedge [1]
Model signs: in-class, out-of-class	Signs taken from the set of model signs, averaged signs [2]
Single signs of the original cuneiform	Averaged sign taken from the original cuneiform [3], [4], [5]
Complex input scene taken from the original cuneiform	Averaged sign taken from the original cuneiform
Complex input scene taken from the hologram of the original cuneiform	Averaged sign taken from the original cuneiform or the hologram

As a result of our experiments we have found that the Multifunctional Extended Optical Correlator (MEOC) gives more extended experimental possibilities by two additional filter planes for filtering in the Fourier and the image plane. The influence of the input aperture is minimized and the correlation results are of higher quality by the multiple prefiltering possibilities. Because of the application of a photoplate for the production of a hologram as matched spatial filter, it is relatively sensitive for vibrations in its surroundings. On the other hand, the Joint Transform Correlator provides the possibility to carry out the correlation experiment in real time because of the application of liquid crystal displays. The input and the reference signal are in the same plane, resulting in a lower vibration sensitivity.

3 Experimental Results

The result of the correlation experiment is shown as an example for the application of the MEOC with an averaged matched filter for the cuneiform sign I-av, taken from the cuneiform tablet HS 158b, Hilprecht collection, Jena (Fig. 1).

Fig. 1a Optically averaged cuneiform sign I-av

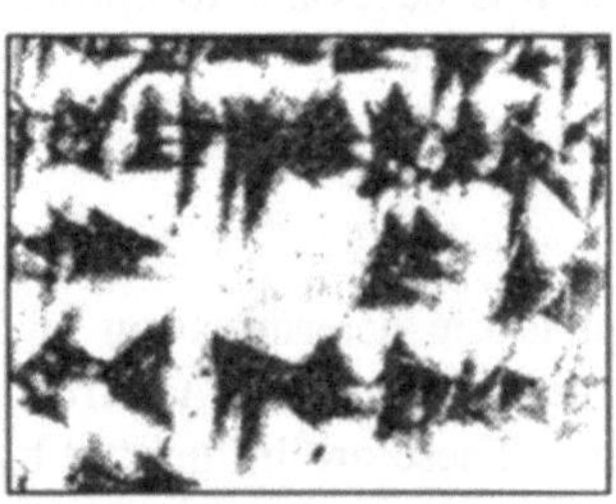

Fig 1b Original cuneiform scene with sign I in the center

Fig. 1c Correlation signal, indicating the presence of sign I in the scene of Fig. 1b

Different averaging methods (Multiplexing, optical and digital averaging, respectively) gave comparable results; the recognition sensitivity is significantly higher in comparison to the application of single sign filters. The method of digital averaging seems to be more easily applicable for a larger number of samples. The application of the MEOC is relatively complicated because of the preprocessing steps, which are necessary for the production of the holographic filter.
The comparison of a scene, taken from a hologram of HS 158b, with an averaged sign is shown as an example for the application of the Joint Transform Correlator in Fig.2.

Fig. 2a Cuneiform scene with sign I and reference sign I-av

Fig. 2b Correlation signal of the JTC

4 Conclusions

The experiments have shown the applicability of methods of coherent optical signal processing in cuneiform research. The correlation of single signs and object scenes in cuneiform tablets was demonstrated with the MEOC and JTC. The application of prefiltering methods gives a higher sensitivity of the correlation experiment. Filters produced from single signs can give misclassifications. The application of averaged correlation filters gives a lower in-class sensitivity and a better out-of-class discrimination. A higher signal-to-noise-ratio is possible by digital preprocessing. The application of liquid cristal displays in a Joint Transform Correlator gives the possibility of a real-time comparison of a cuneiform scene with an averaged cuneiform sign.
As a result, it is possible to decide without previous knowledge, if fragments are written by the same writer or the the same writing school.

Our work was supported by the German Ministry for Education, Science, Research and Technology (BMBF) under contract No. 03WE9HUB.

References

1 N. Demoli, H. Gruber, U. Dahms, G. Wernicke, Journal of Modern Optics **42**(1) 191-195 (1995)

2 G. Wernicke, N. Demoli, U. Dahms, H. Gruber, Naturwissenschaften **82**(9) 395-402 (1995)

3 H. Gruber, G.Wernicke, N. Demoli, U. Dahms, in *Holography and Correlation Optics II*, Chernovtsy 1995, O.V.Angelsky (Ed.), SPIE-Proceedings 2647, 138-144

4 N. Demoli, U. Dahms, B. Haage, H. Gruber, and G. Wernicke, Fizika **A 4** (3), 581-590 (1995)

5 U. Dahms, B. Haage, H. Gruber, G. Wernicke, N. Demoli, Optik **101** (1996) (in press)

Digital Character Recognition of Cuneiform Inscriptions via Neural Networks

A. Roshop*, C. Cruse**

* Laboratory of Biophysics, Institute of Experimental Audiology, University of Münster, Robert Koch Street 45, D-48147 Münster, Germany
** Institute of Applied Physics, University of Münster, D-48147 Münster, Germany

Abstract. Automatic recognition of complex symbols is demonstrated by the deciphering of ancient inscriptions. We show that artificial neural networks can be a useful method to solve this task.

1 Introduction

The deciphering of old inscriptions is an important goal for the study of ancient cultures because these inscriptions contain much information about trade, politics and lifestyle [6]. The most typical method for working with deciphering consists of the following two parts:

1. Transformation of the original symbols into standard symbols, which can be easily read
2. Translation of the standard symbols into the desired language, e.g. modern English

Topic 1. is in particular time consuming, boring and dependent upon the subjective interpretation of the archaeologist. Therefore an automatisation of the transformation of the original symbols into standard symbols would be a great advantage.

Old inscriptions are a kind of handwriting. This presents problems for automatisation because the possible variations of one character in shape make it difficult to lay down exact rules for its identification. Neural networks have proved to be a useful method for the identification of modern handwriting [3][5]. So our idea is to use this method for the identification of ancient inscriptions.

As an example, we investigated cuneiform on clay tablets. This ancient inscription was invented nearly 3000 years before Christ in Southern Iraq. The cuneiform was written by scratching a stylus, made of wood or reed, onto damp clay [6]. Thousands of these clay tablets have been excavated, but most of them have not been translated or even published until now.

Because the originals are stored in museums or collections and can in many cases not be lent out, we used hologramms of the three dimenional cuneiform tablets as information carriers for documentation. The following automatic character recognition was done by the reconstruction of the holographic stored image of a cuneiform tablet.

2 Method

It is widely known that the human brain is able to handle even difficult tasks of pattern recognition. This is the reason why a lot of research is directed toward simulating the abilities of the brain by computer [1][7].

We used a neural network consisting of an input layer, one hidden layer and an output layer. Each neuron of one layer was connected to all neurons of the following layer (see figure 1). The strength of the connection between two neurons can be adjusted by a weighting factor. By choosing suitable weighting factors, the network can distinguish between the different input signals. For finding the fitting weighting factors the network was optimized via examples.

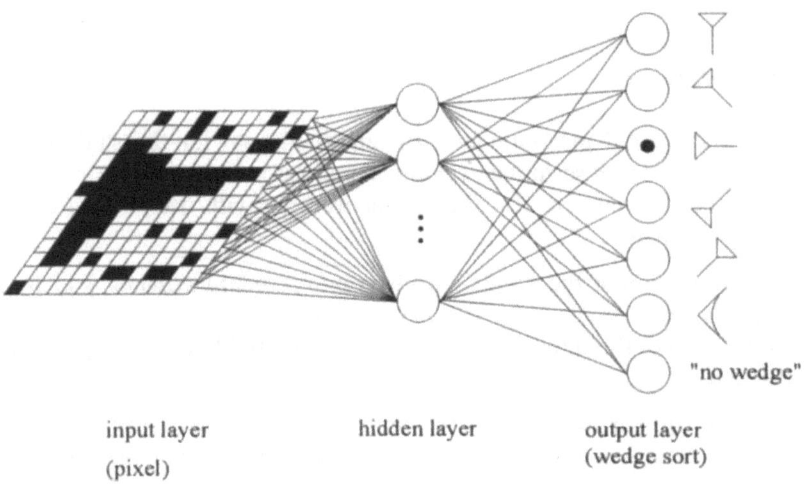

Fig. 1. The pixel information of a part of the digitized image was used as input data. These data travel through the hidden layer of the neural network to the output layer. The output neurons correspond to the possible cuneiform signs. In this example output neuron 3 is activated.

The training of the neural network for cuneiform recognition was done in the following way:

1. examples of wedges and counterexamples were selected
2. the neural network was trained by Backpropagation and the Mikado-algorithm [2]
3. test of the neural network

4. if the classification does not satisfy:
   – select additional examples and recur to step 2.
   or:
   – training complete

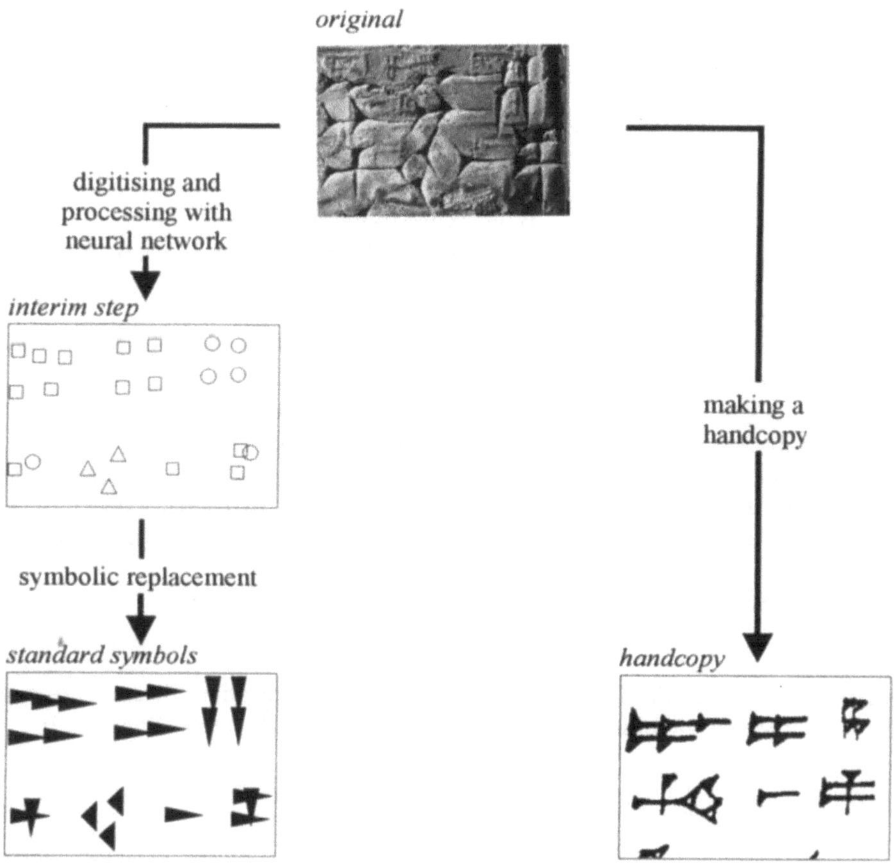

Fig. 2. Usually the archaeologist manufactures a handcopy of the cuneiform inscription (see lower right picture). For the automatic character recognition via neural networks the original tablet was at first digitized. After training, the positions of the wedges were detected (upper left picture) and replaced by the corresponding standard symbols (lower left picture). The translation of the text is: 9.4 "gur" of grain are lent for a rent of 20% per "gur".

3 Results

The study showed that neural networks are an appropriate method for the recognition of handwritings like cuneiform inscriptions, resulting in the following advantages:

1. The faster evaluation time means that a larger amount of cuneiform tablets can be transformed.
2. This task can also be done by non archaeologists.
3. Objective tranformation of the inscription is independent of the knowledge and the subjective interpretation of the archaeologist.

References

1 R. Brause "Neuronale Netze", B. G. Teubner-Verlag, 1995

2 C. Cruse, S. Leppelmann, A. Burwick, M. Bode "The Mikado Algorithm: Dimension Reduction and Training Set Selection", Proceedings of the EUFIT 96, Aachen, September 1996

3 J. Hertz, A. Krogh, R. G. Palmer "Introduction to the theory of neural computation", Addison Wesley, 1994

4 T. Y. Kwok, D. Y. Yeung "Constructive feedforward neural network for regression problems", Technical Report, HKUST-CS95-43, 1995

5 U. Kreßel, J. Schürmann, J. Franke "Optical character recognition and pattern classification", Neuronale Netze - VDI-Berichte 1184, VDI-Verlag, 1995

6 C.B.F. Walker "Cuneiform", University of California Press / British Museum 1987

7 A. Zell "Simulation neuronaler Netze", Addison Wesley, 1994

Financial support of the German Federal Ministry of Education, Science, Research and Technology is gratefully acknowledged (Contract number: 03-VB9MUE-9)

The Use of Digital Image Processing for the Documentation of Monuments and Sites

Jürgen Heckes and Annette Hornschuch

Deutsches Bergbau-Museum, Am Bergbaumuseum 28, D-44791 Bochum

Abstract. Many tasks require the geometrical registration of images before further processing. This can be achieved through digital photogrammetry, whereby the required geometrical input data can be extracted almost exclusively from photos.

1 Introduction

Since 1979 the Department of Photogrammetry within the Deutsches Bergbau-Museum (DBM) has been dealing with the development, modification, and application of modern documentation techniques in support of those sciences taking part in the conservation of cultural heritage.

Apart from the geometrical registration of the realities found, there is an urgent need for scaled photo-realistic representations combined with descriptive information.

In the following chapter, methods of digital image processing with the intention of creating geometrically correct and photo-realistic reproductions of objects will be introduced.

2 Digital Image Processing and Rectification

2.1 Recording

The practical work shows that the optimum combination of recording-systems seems to be that of a medium format metric camera, whose pictures can be interpreted through analytical photogrammetrical processes to acquire highly exact geometrical information, and of a large format system-camera with adjustable lens- and film-units. Using this equipment, almost all terrestrial photograph arrangements aiming at complete object representation can be realized.

2.2 Scanning

Conventional evaluation methods using analytical plotters allow -depending on the interpreter's experience level- virtually any enlargement factors up to grain resolution of film-material.

In [1] the information capacity of a highly resolving film is defined as $2.9 * 10^6$ Bits / cm^2 which is a scan resoluton of 4325 DPI.

If a reproduction of the findings in the form of a printed map is desired, practical experience has shown that photo-realistic reproductions offering 60 linepairs per cm, which can be compared to a resolution value of 300 DPI, do suffice when viewed by the naked eye. This means that images on high-resolution film can be enlarged up to 14 times under optimum conditions.

2.3 Image Enhancement

Image enhancement refers to the process which helps human users with interpreting and supplies an excellent basis for further steps in image processing.

For this process, numerous procedures have been developed which, according to the respective task, change the intensity values of the initial pictures while conserving the picture coordinates at the same time. Two operators, the shading correction and the histogram matching are particularly suitable for the pictorial and geometrical representation of objects.

The shading correction, with locally varying background signals, rectifies different illumination influences, e.g. caused by loss of brightness towards the edges in wide-angle lenses and/or one sided illumination of the object and the histogram matching updates the grey levels in different images.

These pre-processing steps facilitate the automatic matching of image characteristics in stereophotographs for the calculation of altitude models and for piecing together adjoining pictures in a map.

2.4 Orthorectification

"Orthorectification is the process of removing the effects of image perspective geometry and additionally the elevation effects from an image" [2].

Depending on the depth structure of an object's surface and on what degree of exactness must be expected, rectification can be achieved through a multiple-image or single-image process.

Multiple Image Rectification This type of transformation can be applied to objects with any kind of surface geometry.

In this case, interior and exterior orientation of images, as well as digital surface models, are needed as starting parameters for transformation.

The parameters of interior orientation are obtained through calibration of the recording system.

The exterior orientation of images (position of projection centers and of camera axis) is achieved by transformation of homologue pixels onto control points which have been calculated geodetically.

The surface model is defined by any 3-dimensionally defined points and by polygones defining breaklines and interpolation limits. This generally extremely time-consuming process can be automated by methods of digital image processing. In this field, numerous methods have been developed which look for corresponding points in stereoscopic pairs of images in order to calculate the height of the object via the relative position of these points towards each other (parallaxes). This process is called matching or correlation, and it compares and localizes intensity values while at the same time including their surroundings in images.

Matching-algorithms are used successfully in remote sensing applications. Their application in evaluating facades with depth stuctures within the framework of "mass-tasks" has failed so far due to typical surface characteristics such as extreme contrasts, abrupt changes in curvature and hidden areas. Here, further studies are most urgently required, which the Deptartment of Photogrammetry within DBM supports by supplying places for post graduate studies whose results will be published in due time.

Single Image Rectification This type of reshaping, used for approximately plain objects, can be performed on the basis of relatively little initial information.

Figure 1 shows on the left side two original photos of a segment of the side chapel of St. Marienkirche in Herzberg / FRG and on the right side the orthoimage, calculated by a single image rectification process, of the same segment.
As a pictural basis, either metric or amateur images can be used. For the geometrical fit-in information only few control points are needed, which must be well distributed on the object's surface as x- and y-coordinates. These points are fitted into the pictorial coordinate system by measuring the fit-in points in the scanned image, so that parameters for the transformation can be calculated.

Figure 1

In 1992 DBM developed a Unix-based software tool for single image rectification based on a previous analytical photogrammetrical evaluation. For further information see [3].

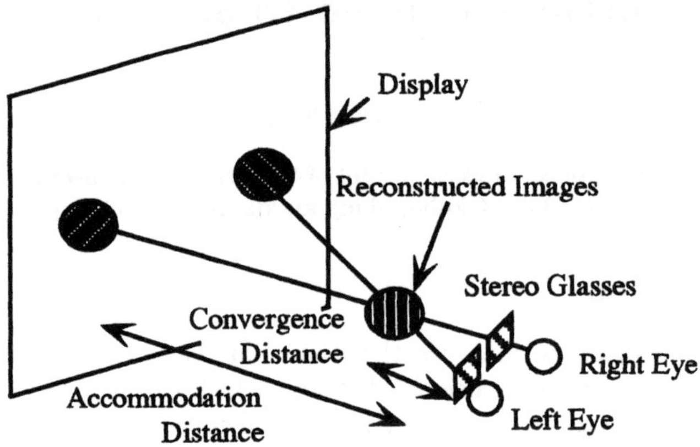

Fig. 1. Mismatch of Distance between Accomodation and Convergence

3 Experimental results

1) Vision functions while looking at 3D images

Experiments were performed with the time-sharing type stereoscopic 3D display using liquid crystal shutter glasses (the alternating rate is 60 Hz). The target 3D image was a white-filled square (1×1 deg. of arc) generated by computer graphics and perceived in front of the display. Accommodation responses were measured with an infrared optometer and convergence responses were measured with an Eye Mark Recorder. Fig. 2 shows an example of the data regarding the changes in accommodation and convergence with a viewing distance of 1 meter as a function of the stereoscopic distance of the 3D images from the display plane. Here, a measure "D" (Diopter) which is the inverse of the length measured in meters is usually used in expressing distance. If we were to maintain clear vision, accommodation (focus) should remain unchanged at the display plane, regardless of the convergence. However, the experiment shows that accommodation also changes in the same direction as the convergence. These results suggest that convergence causes accommodation changes. Fig. 3 shows the data of another experiment in which accommodation response were compared between the stimuli of real objects and artificial 3D images. When viewing the 3D images, the accommodation responses were clearly smaller than those found when viewing the real objects. Pupil sizes were also measured at the same time, and it was found that they were smaller than when looking at stimuli of the real objects.

These results suggest that stereoscopic 3D images with large binocular disparities cause some confusion in the human visual system when viewed. These facts may stem from the close relationship among the accommodation, convergence and pupil size response of the human visual system, probably interactively processed in the higher level of the brain, although this has not yet been clarified.

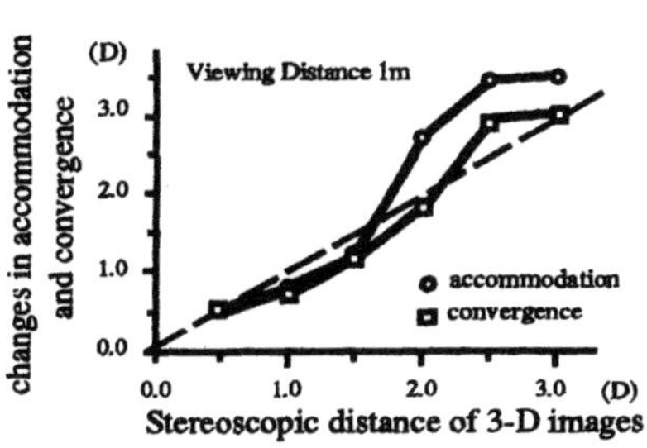

Fig. 2 Changes in Accomodation and Convergence to 3-D Images

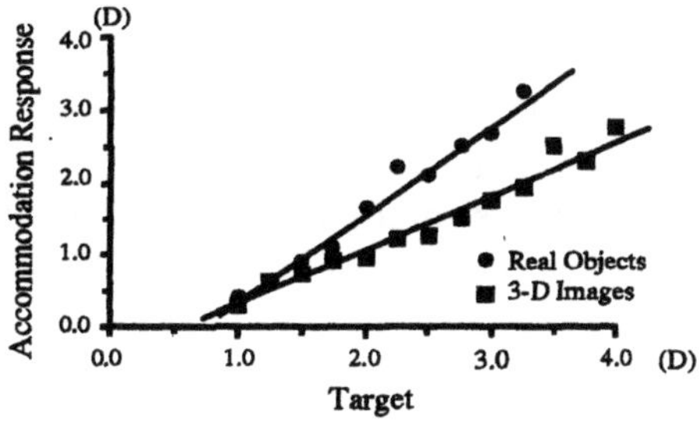

Fig. 3 Accomodation Response to Real Objects and 3-D Images

2) Estimation of eye fatigue after viewing stereoscopic 3D images

Fig. 4 shows another serious experimental result of far-to-near accommodation response time to a step displacement of a target, measured (a) before and (b) long after(60 min.) viewing stereoscopic 3D images, respectively. It was observed that response time becomes longer, and the accommodation cannot maintain the tonic state. This result suggests that beyond some limited range of disparity, the influence of artificial 3D images becomes a serious problem. On the other hand, in the case of normal real images, it was observed that there are no significant changes of response time to the displacement stimuli and tonic state between before and long after looking at the images.

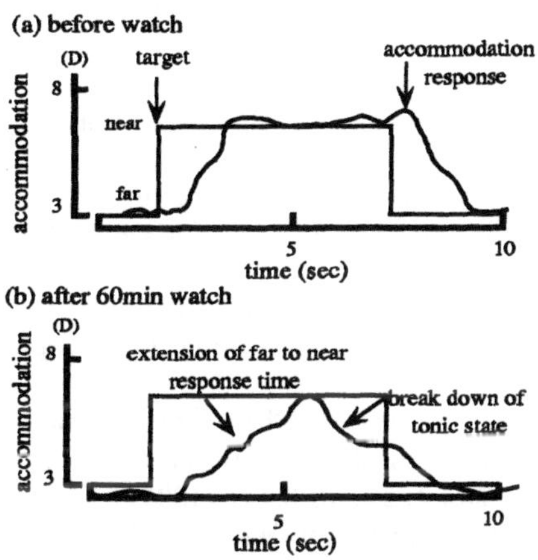

Fig. 4. Changes of Accommodative Function After Viewing 3-D Images

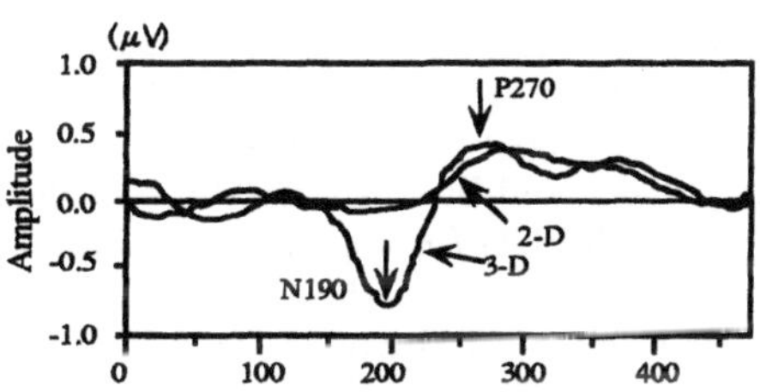

Fig. 5. VEPs Related to Stereoscopic 3-D Images (at Oz: Afterbrain)

Such imbalance of vision functions may be one of the reasons for the visual fatigue associated with artificial stereoscopic - 3D images. Fatigue is both a subjective sensation and an abstract concept which is not measurable by objective methods but may, at least, correlate with decrease, change or deviation from normal visual functions such as accommodation, vergence. pupil size, eve-movement and critical flicker fusion frequencies, etc.

Visual Evoked Potentials (VEP) measured at Oz (afterbrain) of the International 10-20 System show the characteristic two peaks which might be related to binocular vision. These amplitudes increase with large disparities of binocular stereoscopic stimuli, also the latency changes related the displaying time. An example is shown in Fig. 5 where negative N 190 and positive P 270 peaks are the typical signal for 3D images.

4 Acknowledgments

The experiments described briefly in this paper were performed in my laboratory mainly from 1987 to 1991. The author wishes to thank Assistant Professor T. Inoue for leading the studies and the other student members for their efforts.

References

1. Ohzu, Visual perception mechanism related binocular stereo vision technology (in Japanese), Joumal of the Japan Society of Precision Engineering, Vol.54, No.2, pp.225-258 (1988)
2. T. Iwaki and H. Ohzu, 3D Display System and Visual Evoked Potentials (in Japanese), Journal of Ophtalmological Optics Society of Japan, Vol. 10, No. 1, pp. 183 -187 (1989)
3. T. Inoue and H. Ohzu, Eye-Movement and Accommodation while Watching Stereoscopic TV(I) (in Japanese), Journal of Ophtalmological Optics Society of Japan, Vol. 10, No. 1, pp.208-211, (1989)
4. T. Inoue and H. Ohzu, Measurement of the human factors of 3D images on a large screen, in Large-Screen Projection Display II, Proceedings of the SPIE, Vol. 1255, pp. 104-107 (1990)
5. T. Inoue and H. Ohzu, Accommodation and Convergence When Looking at Binocular 3D Images, in Human Factors in Organizational Design and Management III, Elsevier, pp.249- 252 (1990)
6. T. Inoue and H. Ohzu, Eye-Movement and Accommodation while Watching Stereoscopic TV(II) (in Japanese), Journal of Ophtalmological Optics Society of Japan, Vol. 11, No. 1, pp. 159-164, (1990)
7. K. Kagitani, T. Inoue, S. Murakami, K. Yanashima and H. Ohzu, Effect of Eccentricity on VEPs to Binocular Stereoscopic Stimuli (in Japanese), Journal of Ophtalmological Optics Society of Japan, Vol. 12, No. 1, pp. 13 8-141 (1991)
8. H. Ohzu, Artificial 3D displays and visual function, Technical Digest of OSA Topical Meeting 2, Santa Fe, NM., Feb. pp.48-5l (1994)
9. H. Ohzu and K. Habara, Behind the Scenes of Virtual Reality : Vision und Motion, Proc. IEEE, 84, No.5, MAY, pp.782-298 (1996)

Visual Stimulation Techniques at the End of XX Century

Lucia R.Ronchi

formerly Director of Research in Visual Science
at the Istituto Nazionale di Ottica,
Arcetri 50125, Florence, Italy

Abstract- The versatility of computer generated stimuli on phosphor displays has been flanking the advancement of Visual Science, but an end point is being reached: the intrinsic technical limitations fail in competing with the fineness of the "psychophysical evaluation of image quality" in the spatio-temporal realm.

1 Introduction

In the classical visual research covered by the umbrella of "early vision", all variables but one were "frosted" and the basic visual functions were thus recorded. The advanced visual reseach tackles, in a spatio-temporal realm, with moving and 3D patterns, test-fields embedded in crowded, complex surroundings, the interactions of the outputs of various "channels", in turn processed by simple, complex and hypercomplex "units", in harmony with the paradigm of "tuning".

As Daugman [1] notes, there is a basic dichotomy in the research on early vision; the space-domain local feature detection and the Fourier-like decomposition into spatial frequency components.

Daugman proposes an extension to the visual process of Gabor's Communication Theory, including the indeterminacy relations of Quantum Mechanics. This leads to an expression of the theoretical limit of joint 2D resolution in the 2D domain. As a further extension, the non-Fourier analysis is nowadays tackled once the linearity-non linearity intricacies have been solved.

2. The escalation of complexity of visual stimulation as deduced by sampling visual journals

In harmony with the present space allowance, let us limit ourselves to compare the situation in the mid eighties to that in the mid nineties. In the former case, one finds that the Optical Devices are used with the same frequency of occurrence as the CRT stimulators (except for transient aspects of color vision, where Optical Devices prevailed). Visual stimuli used may be subdivided as follows:

a)- the traditional stimuli, heirs of the cut-and-patch technique, static or presented under time-varying conditions;

b)- the luminance sine wave patterns, modulated in amplitude and/or spatial frequency, static, or in motion, with variable orientation, in the spatio-temporal realm where phase and frequencies are manipulated as ingredients of the language of Fourier analysis, in the framework of selectivity of adaptation and masking;

c)- an escalation towards complexity, through stationary compound gratings or with components moving in various directions, the elimination of the fundamental frequency, the counterphase modulated gratings, and through the interaction between test and background and/or context complexity;

d)- the spatio-temporal derivative scheme, in 1D and 2D situations, from the Gaussian, to the difference of Gaussians (DOG), the 6th derivative,(D6), and the the 10th derivative (D10) of Gaussians, the spatio-temporal "windowing", the Gabors, by manipulating the bandwidth , the size and the blur. In particular, such stimuli are of vital importance when used dichoptically, with different disparity, that is, as "stereopairs", static or in motion;

e)- the random dots, static or moving, as test stimuli, per se', or as constituents of patterns, or as noise, or as random dot kinematograms;

f)- patterns sequentially presented, intervalled by blanks, by other targets differing in orientation, disparity, noise, etc.

In the mid nineties one finds two major facts, all rendered possible by the prevailing use of computer-generated stimuli, compared to Optical Devices:

1)- situations where various visual functions interact, by the use of multidimensional stimulations, in view of items such as "segregation", "segmentation","clustering", and similar;

2)- the use of targets, stationary or in motion, tuned to the response profile of specific visual units, such as: moving plaids; compound Gabors; bi-dimensionally and spatio-temporally windowed; complex textures; random dots so modulated as to bypass the constraints of early vision, thus allowing the study of higher order visual processing, as related to a new interpretation.

In other words, the number of dimensions of a visual stimulus is thus increased. For instance, it is defined by **four contrasts:** luminance, color, texture, stereodisparity.

3. The limitations of computer generation on phosphor displays

The problem is complex. Here we limit ourselves to a few basic examples:

3.1 Spatial aspects

The fine positioning of a pattern element is of vital importance in some experiments (for instance, those on perturbations of hyperacuities, where distances ranging from a few min of arc to even 1 sec of arc are implied). The use of a viewing distance ranging from, say, 2.5 to 8 m is a questionable solution. Algorithms aiming at optimizimg sub-pixel accuracy are now appearing in the literature.

3.2 Temporal aspects

The interaction between the phosphor persistence and the visual persistence dramatically limits the predictability of temporal patterning of stimulation. The photometric assessment is only one of the items. The other is represented by the psychophysical assessment , the guidelines of which are still a matter of discussion.

3.3 Transient color stimuli

Some experts in color vision assert that the dependencies of rise time of the primaries of an RGB monitor are so complicated, that the assessment of the temporal patterning of a transient stimulus cannot reliably support quantitative assessment.

3.4 The prediction of appearance

In steady state situations, the prediction of appearance is strongly limited by the fact that the stimuli generated on a phosphor screen are self-luminous. Some phenomena like, for instance, perceptual color constancy of known objects require the perception of "surface" or "object", which is not easy to be achieved, even if the pattern is viewed on or surrounded by a background of the proper luminance. Observer's criterion and attitude are not so easy to be manipulated by the experimenter as desired by crude modeling. The "reference", of basic importance in the above framework, is assumed to be present on the display, or in the environment, or belongs to the previous experience of the observer. In this connection, let us recall that after the Rea, Robertson, Petrusic [2] effect, the skin of individual's hands represents a precious and very fine photometric-colorimetric, "portable" reference for the evaluation of appearance. Let us recall that appearance is even influenced if hands are generated on a display, together with the elements of a patterning.

3.5 Psychophysical evaluation of image quality

The presence of pixels on a phosphor display affects the shape of that threshold contour that is named temporal modulation transfer functions (tMTF) in such a complicated way, that the standard observer defined for traditional stimuli does not apply to monitors. We have been proposing to evaluate the quality of the generated image by consider flicker conspicuity at suprathreshold levels [3] according to an eight-point subjective scale. Now, once the device is fixed, one can extend the above evaluation to the quality of samples reproduced on a transparent film and applied on the display, on which an intermittent stimulus is generated. This is the case, for instance, in the degradation of the subsequent photocopies of a given printed page. The interaction of the pixel structure with the progressively poorer contrast results in a significant change in flicker conspicuity

References

1 J.G. Daugman, J. Opt. Soc. Am. A2, 74 (1994).
2 M.S. Rea, A.R.Robertson and W.M. Petrusic, Color Res. and Appl.15,80 (1980)
3 L.R. Ronchi, J. Photogr. Sci. 42, 74 (1994)

Unconventional Microscopy

Superresolution in Far-Field Microscopy

Christiaan H.F. Velzel[1], H. Paul Urbach[2], Robert Masselink[3]

[1] Philips Center for Manufacturing Technology, P.O. Box 218, 5600 MD Eindhoven, The Netherlands
[2] Philips Research Laboratory, Prof. Holstlaan 4, 5656 AA Eindhoven, The Netherlands
[3] Nanofocus Messtechnik GmbH, Bismarckstrasse 120,47057 Duisburg, Germany

Abstract. The theoretical possibility of reconstructing images of objects with structural details smaller than the defraction limit was shown already many years ago. By scanning near-field optical microscopy this possibility has been realised. We obtained noise-limited resolution in farfield microscopy. In this paper we discuss the experiment and some of its applications.

1 Introduction

Superresolution has been realised in scanning near field optical microscopy [1]. Useful information is obtained by this method, although a general, consistent theory does not seem to exist. In the field of superresolution by far field microscopy, the situation is almost the reverse. Here a general theory exists [2], but very few experimental realisations have been reported. In this paper we show that the existing theory proves the possibility of superresolution, but at the same time hinders its realization. So much so that experiments like those of Tychinsky [3] that cannot be explained easily by the existing theory but nevertheless show superresolution have been met with disbelief and indifference.

We in the following review shortly the existing theory of superresolution and its conclusions as to the formation of superresolution images. We then describe a preliminary theory of our own that is concerned with the formation of superresolution images. Thirdly, we show results of some experiments in superresolving interference microscopy.

There seems to be some uncertainty as to the definition of superresolution. In the now classical papers of Lukosz [4], the resolution of a system with a given numerical aperture is increased by grating modulation of the object field (and demodulation of the image). This results in a decrease of the width of the usable field of view, so that the number of degrees of freedom in the image [5] remains constant. Still the resolution is limited by diffraction in this case. In this paper we define superresolution as the situation in which details in the object structure can be detected that are smaller than half the wavelength, thus exceeding the Rayleigh limit. In this case the resolution is not limited by diffraction, but by noise (coherent noise or detector noise).

2 Theories of Superresolution

It was first shown by Toraldo di Francia [2] that diffraction sets a practical, but not a fundamental limit to resolution. We will come back to his arguments later in this Sect. Here we first discuss an experiment by Wolter [6] that is an early realisation of superresolution and provides us with the simplest example of a theory. Wolter set up a finite row of dipole microwave antennas at distances much smaller than the free propagation wavelength. He measured the complex amplitude in the far field from which he calculated the structure of its source. From his expression for the far field amplitude follows that the reconstructed amplitude in the source plane can be written

$$u(x) = \sum_n a_n \mathrm{sinc}[2\pi \sin(\theta)\,(x - x_n)/\lambda], \tag{1}$$

where a_n is the strength of the n-th dipole and x_n is its postion, λ is the wavelength and $\sin\theta$ is the numerical aperture of the far field. When the positions of the dipoles satisfy:

$$-\frac{\lambda}{2\sin\theta} < x_n < \frac{\lambda}{2\sin\theta}, \tag{2}$$

so that all dipoles lie within the central maximum of the diffaction spot, it is possible to calculate the amplitudes a_n from a sufficient number of measurements of $u(x)$ inside this interval. Wolter calculates $u(x)$ from the far field data; in optical microscopy, like in the experiments of Tychinsky [3] and the experiments reported below, the reconstructed field is the field in the image plane of a lens, at a high enough magnification so that a direct measurement of $u(x)$ is possible.

It is seen from (1) that the information about the source (the coefficients a_n), is present in the reconstructed field $u(x)$, but in a scrambled form. Because the sinc function does not change much over a distance small to the width of its central maximum, the reconstruction of the source from $u(x)$ must be done by high precision calculation. The presence of noise in the far field measurements leads to degradation of the reconstruction (dependent on the type of noise). This point is discussed in Wolter's paper [6] in which he showed that a moderate amount of phase noise does not degrade the reconstruction seriously. In this way he was able to reconstruct an object consisting of dipoles at a distance of 0.1 λ. It is clear that the number of dipoles that can be resolved within the limits of (2) depends on the signal-to-noise ratio obtained in the experiment. This agrees with our definition of superresolution.

Summarizing, two things are essential in Wolter's experiment: that the object is small (of the order of the wavelength) and that its structure (a row of dipoles with given distances) is known a priori. Both issues have been discussed in the literature [7], [2] so that we do not dwell on them any further.

A theory of superresolution valid for continuous objects was presented by Tiraldo di Francia [2]. This theory is based upon the properties of prolate spheroidal functions as described by Slepian and Pollack [8]. For further developments of this theory, see Bertero and Pike [9], and de Mol [10]. We give a summary of the essentials of this theory, for the case of coherent illumination. The relation between the object amplitude $v(x)$ and the image amplitude $u(x)$ is given by

$$u(x) = \int v(x')\mathrm{sinc}[2\pi \sin(\theta)(x - x')/\lambda]\, \mathrm{d}x'. \tag{3}$$

Equation (1) corresponds to the discretized integral. By expanding $v(x)$ in terms of the eigenfunctions $g_i(x)$ of the operator acting on $v(x)$ in (3) — convolution by a sinc function — so that

$$v(x) = \sum_{i=1}^{\infty} a_i g_i(x), \tag{4}$$

we obtain

$$u(x) = \sum_{i=1}^{\infty} \lambda_i a_i g_i(x), \tag{5}$$

where λ_i is the eigenvalue belonging to eigenfunction g_i. Again, the object coefficients a_i can be found from measurements of $u(x)$. The eigenfunctions $g_i(x)$ turn out to be the linear prolate spheroidal functions [13]. Their eigenvalues depend in a significant way on the size of the object domain. We take over Fig. 1 from the paper of Slepian and Pollack [8]. There the eigenvalues are given for a number of values of i as functions of the classical number of degrees of freedom [5], [9], given by

$$S = \frac{2\sin\theta}{\lambda}X, \tag{6}$$

where X is the size of the object domain. It is seen from this Fig. that for values of i larger than S, the eigenvalues go to zero rapidly with increasing i. This behaviour is strongest for large values of S, $S \gg 1$; in this way the theory of resolution of Abbe is corroborated. For values of S near 1, the decrease of λ_i with i is slower, showing that superresolution is realised easier with objects of the order of the wavelength. From (2) we see that in Wolter's experiment we have $S \leq 2$. The consequences for image reconstruction can be described as follows. The number of degrees of freedom in the object field is infinite, according to (4). The number of degrees of freedom in the image field is infinite in principle, as we see from (5). In practice this number is limited by the condition that $\lambda_i a_i$ is sufficiently larger than the noise amplitude, so that we have

$$u(x) = \sum_{i=1}^{K} \lambda_i a_i g_i(x), \tag{7}$$

where K is of the order of S. This question is treated in Bertero and Pike [9].

In the next Sect. we show pictures of phase objects obtained by interference microscopy with a large magnification, so that the CCD detector used can resolve details in the image field that are much smaller than the Airy disc. Under these circumstances we obtain images that correspond remarkably well to the object. Such images cannot be explained very well by the theory of superresolution

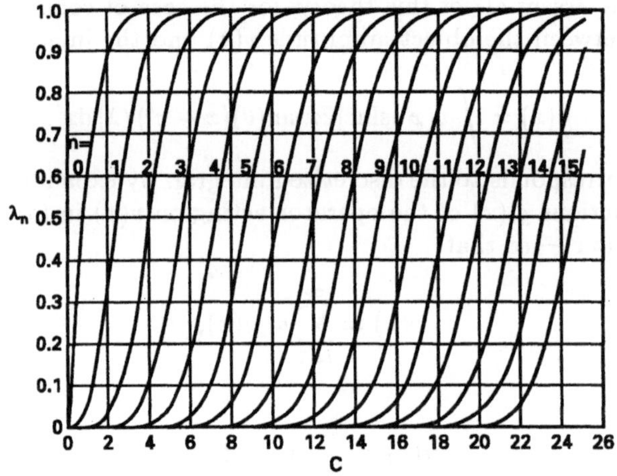

Fig. 1. Eigenvalues of linear prolate spheroidal functions (from [8]).

described above. According to this theory, as summarized in (7), only those objects that consist of eigenfunctions $g_i(x)$ with the same eigenvalues λ_i are reproduced identically in the image. From Fig. 1 we see that we can synthesize such objects from object eigenfunctions that belong to different values of S; such an object will have only small details in that part of the field that belongs to the smallest value of S used. Thus far we considered only one dimensional objects. In two dimensions the set of objects that are imaged identically is somewhat larger. For instance all objects of the form

$$v(x, y) = a_k g_k(x) + b_k g_k(y), \tag{8}$$

are imaged identically when the eigenfunctions belong to the same eigenvalue of S. More generally, objects of the form

$$v(x, y) = a_k g_k(x, S_1) + b_j g_j(y, S_2), \tag{9}$$

are imaged identically when $\lambda_k(S_1) = \lambda_j(S_2)$. When we relax the criterium that the object be reproduced identically, we have a greater choice of objects. For instance, when we require that only those points of the object are reproduced where the modulus of v is zero, we can take $g_k(x)$ and $g_j(y)$ in (9) with different values of λ. This has a direct connection to optical phase singularities. Suppose that the real and imaginary parts of $v(x, y)$ are given by

$$\begin{aligned} \operatorname{Re} v(x, y) &= a_k g_k(x), \\ \operatorname{Im} v(x, y) &= b_j g_j(y), \end{aligned} \tag{10}$$

then at those points where $g_k(x) = g_j(y) = 0$ the phase is undetermined. These points are called phase singularities by Bazhenov et al. [11]. According to (8) these singularities are reproduced in the image field. Equation (10) can be extended as follows; when

$$\operatorname{Re} v(x,y) = \sum_k a_k g_k(x - x_k),$$
$$\operatorname{Im} v(x,y) = \sum_k b_k g_k(y - y_k), \tag{11}$$

where the shifts x_k, y_k are chosen so that the eigenfunctions in the sum have a common zero, then the phase singularity at this point is reproduced in the image field. Also the zero contours $\operatorname{Re} v = 0$ and $\operatorname{Im} v = 0$ are reproduced identically. In the examples given above we have taken the contours $\operatorname{Re} v = 0$ and $\operatorname{Im} v = 0$ perpendicular to each other; in a complete two dimensional theory this restriction would not be necessary. Nevertheless, the examples given above show that identical reproduction of objects or details of objects (such as phase singularities or zero contours) is possible, but at the cost of a severe reduction of the number of degrees of freedom. We will come back to this question in Sect. 4.

3 Experiments in Interference Microscopy

A scheme of the experimental set-up with which we performed some experiments on superresolution in interference microscopy is shown in Fig. 2. As a basis for these experiments we took a Nomarski microscope (produced by Leitz in Wetzlar, Germany) that is designed for the study of reflecting phase objects by shearing microscopy. This microscope is equipped with illumination optics with a white light source, which we use to find the correct focus and to select the detail in the object that we want to study. The white light illumination optics is not shown in the Fig.. Using objectives corrected for infinity and a tubus lens we can achieve magnifications up to 250×.

We made two notable additions to the microscope that are indicated in Fig. 2. We added a laser illuminator, consisting of a semiconductor diode laser ($\lambda = 675\,\mathrm{nm}$), collimating optics and a polarizer. The polarization direction was chosen at 45° with respect to the crystal axes of the Wollaston prism, in the microscope, so that the two perpendicularly polarized waves that illuminate the object have equal amplitudes. In the parallel laser beam that is directed by a neutral beamsplitter to the objective we have placed an electro-optical modulator that produces a variable phase difference between the two polarization components of the illuminating wave. The illuminating wave is focused on the object by the objective lens, so that only a small spot on the object , with a diameter of the order of $\lambda/\sin u$ ($\sin u$ is the numerical aperture of the objective) is illuminated. As we have seen in Sect. 3 this is a favourable condition for superresolution. Due to the angular splitting (approximately 17 arcsec.) by the Wollaston prism the polarization components of the illuminating wave are shifted with respect to each other in the direction perpendicular to the crystal

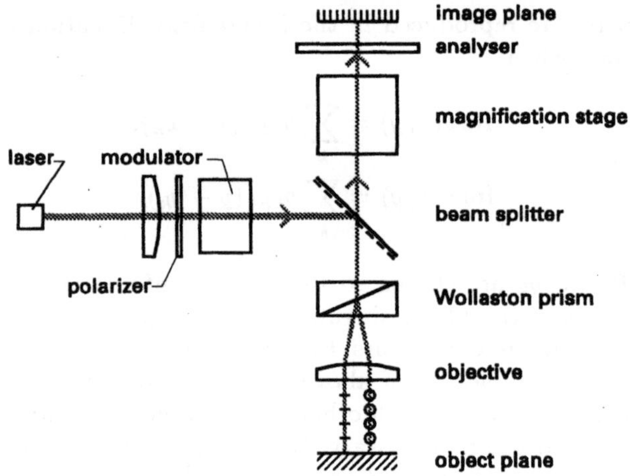

Fig. 2. Experimental set-up.

axes of the Wollaston. With an objective having a focal distance of 4 mm this splitting is equal to 0.3 μm.

Interference between the two polarization components is brought about by a diagonal analyser below the image plane. A second addition to the microscope is a secondary magnification stage (40×) that images the diffraction spot on a CCD detector. In the primary image the diameter of the diffraction spot is about 50 μm, on the CCD detector this becomes 2 mm; this corresponds to about 200 pixels of the CCD detector. With a 50× objective one pixel then corresponds to about 5 nm on the object; with a special objective of 250× we can decrease this to 1 nm. For the time being, the pixel size does not form a limit to resolution.

The phase difference between the polarization components in the image field is measured pixel by pixel by introducing phase differences of 90°, 180°, 270°, 360° between the polarization components of the illuminating laser beam and processing the corresponding CCD images according to a four-bucket algorithm [12] (the image processing software was supplied by Dr. B. Breuckmann GmbH, Meersburg).

We present two phase images made by the set-up we described above. The first of these images is shown in Fig. 3. It is a phase image of a line structure (photo-resist on silicon, covered with a thin gold layer) about 100 nm wide. The linewidth measurement was checked by electron microscopy and AFM profile measurement [13]. The Fig. shows a pseudo-3D image and a cross-section. The deviations from straightness in the pseudo-3D image are caused presumably by a neighbouring structure perpendicular to this line structure. Conspicious in the cross-section picture are the steep slopes of the phase image, corresponding to an edge width of 3 pixels, 15 nm on the object.

The second phase image is shown in Fig. 4. The target consisted of small holes in a silicon surface. Independent measurements gave the distance between

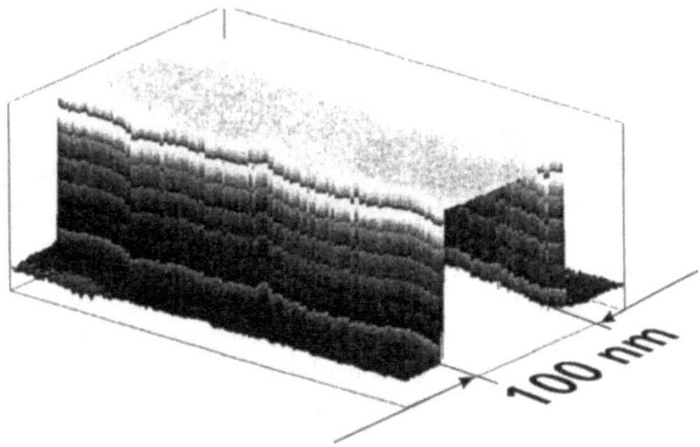

Fig. 3. Pseudo-3D profile of a 0.1 μm wide line.

the centres C and C' of two holes as 0.2 μm. The diameter of a single hole is somewhat smaller than 0.1 μm. In the phase image the number of holes is twice the number of holes in the target. Half of the number of dots shown in the phase image must be artefacts. Again, the edges of the phase images are very steep (of the order of 10 nm) as shown in the cross-section in the Fig..

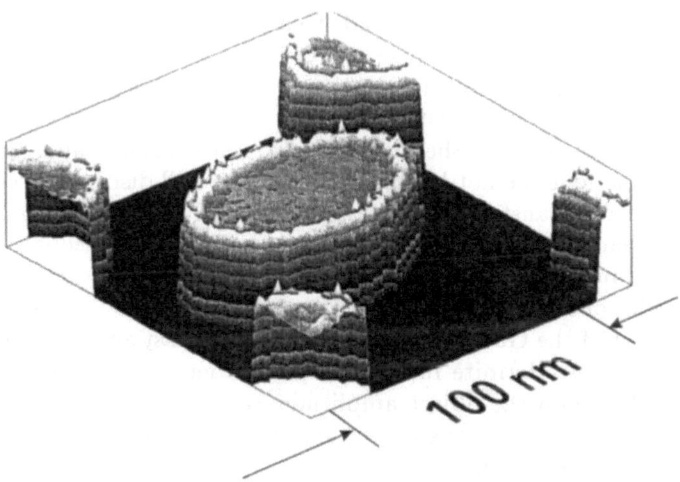

Fig. 4. Pseudo-3D profile of circular structire features.

The experiments presented here were corroborated by other measurements on line and dot structures. They gave rise to the following questions.

1. Does linewidth measured in these phase images correlate with the width of line elements in the object structure?

2. Is the edge width shown in the phase image related to edge steepness in the object, or is it an artefact?
3. Two dimensional, periodic structures seem to give rise to artefacts (e.g. frequency doubling). What are the artefacts of more general structures?
4. Why are the phase images that we obtain at magnifications of $1000-10.000\times$ so remarkably like the object structures?

To resolve these and similar questions, the scattering of polarized light at submicron surface structures must be studied in more detail. Such a study should result in the relationship between the object structure and the amplitude of the field in the object plane. We do not consider this problem in the present paper; in the next Sect. we discuss further relations between the object amplitude and the image amplitude, continuing the discussion from Sect. 3.

4 Superresolution and Image Formation

In Sec. 3 we discussed the theory of superresolution on the basis of bandlimited eigenfunctions of the optical system. We saw that in general the image amplitude resembles the object amplitude only as far as are concerned details of the order of magnitude of the diffraction spot, and that the information about smaller details is weakened so that a scrambled picture of the object amplitude results.

We saw also in that Sect. that the requirement of identical reproduction of the object amplitude in the image leads to a severe loss of degrees of freedom in the object; relaxing this criterium so that only certain features of the object (zero contours, phase singularities) are reproduced leads to somewhat more freedom, but the increase is not impressive.

In this Sect. we describe the formation of images using Gauss-Hermite functions; these functions are not bandlimited, and we will discuss the consequences of their use for the transfer of optical fields through image forming instruments. On the other hand a Gaussian beam has exactly a value of S equal to 1, because the angular diameter of the far field is equal to λ/w where $2w$ is the waist diameter. The width of a Gaussian-Hermite function of order m is a factor $(m+1)^{1/2}$ larger than that of the Gauss function, both in the waist and in the far field [14], so that using Gauss-Hermite functions up to order m we have $m+1$ degrees of freedom. To obtain the object amplitude we use the following simple model of scattering at the object. The object is illuminated by a wave with amplitude $v_0(x)$; the object amplitude is now given by

$$v(x) = T(x)v_o(x), \tag{12}$$

where $T(x)$ is a transmission coefficient. In this sect. we take a Gaussian illumination wave

$$v_o(x) = \exp(-x^2/2w^2), \tag{13}$$

and we suppose that $T(x)$ is of the form

$$T(x) = \prod_{i=0}^{m} (x - x_i) \exp(-x^2/2w^2). \tag{14}$$

Note that the zeros of the polynomial in T may be complex; in this way we can simulate phase functions. The Gaussian function is added in (14) to prevent $T(x)$ from becoming infinite. The object amplitude is now given by

$$v(x) = \prod_{i=0}^{m} (x - x_i) \exp(-x^2/w^2). \tag{15}$$

This can be written as a sum of Gauss-Hermite functions (with complex coefficients) up to order m. We have put all the degrees of freedom that we have in the zeros of the polynomial. We now pose the questions: can we use our freedom to generate objects with details smaller than the wavelength and do these details survive the transmission through a bandlimited system? Note that an optical system always has a limited frequency transfer band, because the numerical aperture cannot exceed unity (in air, or n in a medium of refractive index n) so that the frequency limit is always lower than $2/\lambda$ (higher frequencies lead to the formation of evanescent waves).

The first part of the above question can be answered in the affirmative. We can simulate small details by making the distance between the zeros arbitrarily small. The second part can be answered only by trying. We calculate the far field of $v(x)$ as given by (15), we then impose limits

$$-\frac{\sin u}{\lambda} < \frac{\sin \theta}{\lambda} < \frac{\sin u}{\lambda}, \tag{16}$$

and consecutively transform back to the near field. To ensure that all our information passes the pupil we take

$$(m+1)\lambda/w < \sin u. \tag{17}$$

As an example we show calculations for $m = 2$, $T(x) = x^2 - \delta^2$, see Fig. 5. In this Fig. we show the minimum value of δ for which the two zeros of $T(x)$ survive as a function of the numerical aperture $\sin u$, more precisely δ_{min}/w is shown as function of $(w \sin u)/\lambda$. We see that for $w = \lambda$ and for a numerical aperture of 1 details larger than 0.01 λ can be transmitted. Because we chose δ real, the phase of $v(x)$ is binary; as long as the zeros of $v(x)$ exist this property of the phase remains but the positions of the phase jumps (zeros) are not equal to those in the object.

The zeros of the polynomial in T may also be chosen complex. If δ is complex, the phase of the image will not be binary but will vary more gradually. In this case, there does not exist a well defined minimum value for the real part of δ for which a phase image ceases to exist; in fact for some complex δ with real part that is considerably smaller than δ_{min} a clear phase image can still be observed.

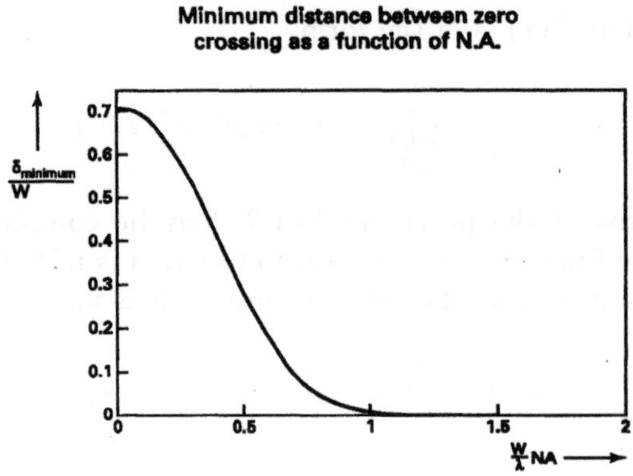

Fig. 5. Minimum distance between zero crossings as a function of numerical aperture.

A phase measurement in the image plane makes sense only when the signal generated by the image field is larger than the noise (caused by stray light or detector fluctuations).

In our example the amplitude at $x = 0$ is equal to δ^2. We assume that the noise is proportional to the maximum object amplitude, which is equal to w^2/e in this case. With coherent detection we find the resolution criterium

$$\frac{\delta^2}{w^2} \geq \frac{e^{-1}}{\text{SNR}}. \tag{18}$$

Such a criterion for noise-limited resolution was also found by Tychinski [15]. We see that in order to resolve details of 0.01 λ, when w is of the order of λ, requires a SNR of the order of 10^4.

Equation (18) is valid for two neighbouring zeros; for an isolated zero the uncertainty of localization is given by

$$\frac{\delta}{w} = \frac{1}{\text{SNR}}. \tag{19}$$

With $T(x)$ an m-th order polynomial,it can be shown that δ/w is proportional to $(1/\text{SNR})^{1/m}$. In the paper of Bertero and Pike [9] also a dependence of resolution on the classical number of degrees of freedom is demonstrated.

5 Discussion

The experiments, those of Tychninski and those reported in this paper, show that noise-limited resolution inside the Airy disc is possible and can be realised in interference microscopy. Our experiments point in the same direction as those of Tychninski, although he used different means to produce them: a Linink interferometer, an image dissector as detector, bootstrap phase modulation and therefore different software and electronics. The classical theory of superresolution is not very helpful in describing superresolving image formation. In this paper we made use of Gauss-Hermite functions to describe the formation of images from object details smaller than the diffraction limit. We explained some of the features of the images found in the experiments, but most of the questions arised in Sect. 3 can only be partially answered at this stage.

Note added in proof: Since the writing of this paper, an article has appeared written by M. Totzeck and H.J. Tiziani (*Optics Communications*, 1996, 4169), in which the authors consider the same subject, using a different approach. Their results show the same tendencies as described by us; but they draw different conclusions.

References

1. A. Lewis, Chap. 17 in *Current Trends in Optics*, Ed. J.C. Dainty, Academic Press, London, 1994.
2. G. Toraldo di Francia, *J. Opt. Soc. Amer.*, **59** (1969), 799.
3. V.I. Tychinski, *Optics Communications*, **74** (1989), 41-45.
4. W. Lukosz, *J. Opt. Soc. Amer.*, **57** (1967), 163.
5. G. Toraldo di Francia, *J. Opt. Soc. Amer.*, **45** (1955), 497-501.
6. H. Wolter, *Arch. Elektr. Übertr.*, **20** (1966), 103-112.
7. C. Pask,, *J. Opt. Soc. Amer.*, **66** (1976), 68.
8. D. Slepian and H.O. Pollack, *Bell System Techn. J.*, **40**, No.1 (1961), 43-63.
9. M. Bertero and E.R. Pike, *Optica Acta*, **29** (1982), 727-746.
10. Christine de Mol, Chap. 17 in *Trends in Optics*, Ed. A. Consortini, Academic Press, London, 1996.
11. I.V. Basistiy, M.S. Soskin and M.V. Vasnetsov, *Optics Communications*, **119** (1995), 604.
12. K. Creath in: *Progress in Optics*, Vol. **26**, 349-393, North-Holland, 1988.
13. J.W. Raith, Dortmund, private communication.
14. A.E. Siegman, *Lasers*, Oxford University Press, 1986.
15. V.I. Tychinski and C.H.F. Velzel, Chap. 18 in *Current Trends in Optics*, Ed. J.C. Dainty, Academic Press, 1994.

Scanning Acoustic Microscopy: Principle and Application in Material Testing of Antique and Contemporary Samples

U. Scheer, K. Kosbi, S. Boseck
Institute of Materials Science and Structure Research, University of Bremen, Germany

Scanning acoustic microscopy (SAM) has proved to be an essential tool in materials science. Ultrasonic waves between 10 MHz and 2 GHz can be used for imaging boundary structures and elastic-mechanical discontinuities within opaque materials, compounds, or biological tissues [1, 2]. It has become an instrument for the development of IC's and microsensors. The lateral resolution on the surface of the specimen can vary with higher frequency, from several µm to 400 nm, while the penetrating depth is reduced from millimeters down to 10 microns in the inner region of a metallic specimen[3]. Here, an introduction to SAM and its special imaging processes is given [4, 5]. Four applications are shown, which are well suited to reveal the inner structures of the specimen.

The principle of SAM

SAM is a confocal microscope, where the focus of an acoustic convergent field is moved in a 2d-scanning pattern over the field of interest on the surface of the sample. Between the acoustic lens and the specimen, a droplet of water is used to couple the ultrasonic wave to the material of the specimen by the longitudinal wave modes within the water. The focus spot and a number of modes of shear waves and longitudinal waves are generated, which are typical for the specimen. We also find modes of surface acoustic waves, which are guided along the border between the bulk and the water droplet.

The ultrasonic wave is produced by an electronic- ultrasonic signal converter on the basis of piecoelectric ceramics (transducer), which is driven by an electronic high frequency signal generator. The acoustic lens is made from a sapphire crystal, showing a negative curvature radius between 25 and 80 µm at the water-side for a working distance and a focal length of the same magnitude.

While in light optics such a lens would suffer from spherical aberration, in acoustic microscopy this can be neglected as the refractive index of sapphire against water (70 °C) for ultrasonic waves is 7.4. Thus, the geometrical imaging defects of a lens are reduced nearly to zero, and a focus spot of one wavelength diameter is produced.

The incoming acoustic signal is reflected partly from the surface, and it is transformed point by point into a monitor signal. This image will show the distribution effects of materials and structures by their elastic discontinuities, but also those from deeper parts of the specimen. Another part of the wave energy may be modified into a surface acoustic wave, which can be rescattered coherently into the lens aperture, thus generating the V(z)-effect of image contrast, which is caused by the coherent interference between the incoming wave within the aperture of the lens and the rejected part. This is an important phenomenon as the phase velocities of the surface wave modes, which are physically generated according to the compartments of the materials or to the ultrasonic wave optical conditions, are clearly related to the elastic constants of the material compartments, its isotropy and distribution over the sample volume and the special kind of compound. The V(z)-effect leads to an alternating contrast within the image when the focus is shifted along the optical axis of the lens. Thus materials can be detected by their elastic-mechanical behavior and the borders of the individual compartments as well as scratches, holes or regions of reduced connection can be seen on a microscopical scale.

The image contrasts in SAM are more complicated than in light optical microscopy as they are combined from several specialized contrast mechanisms:
- V(z)-contrast with alternating sign and amplitude of contrast when the focus position is moved along the optical axis of the lens. It is a nonlinear material contrast.
- Contrast variation in the acoustic impedance of the sample without generation of SAW, especially with small aperture angles of the lens. This is a nonlinear amplitude contrast.
- Contrast variations in the topography of the specimen. For small defocus differences it can be treated like a combination of amplitude and phase contrast and it is possible to correct it by means of linear transfer theory.
- Halo-contrast, which is a microscopic diffraction effect. It is a pure phase contrast and can be corrected by the means of linear transfer theory.

Three examples of investigating opaque samples by ultrasonic waves are shown:

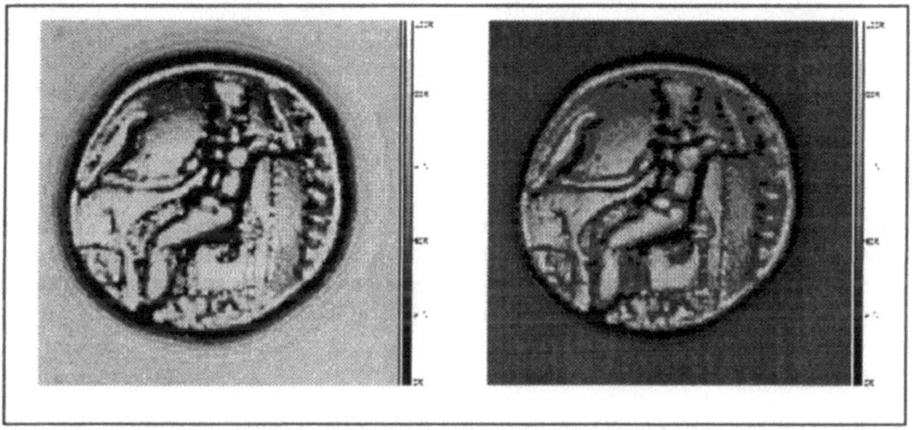

Fig. 1. The two images show acoustic images of an antique Greek coin. In the left image, only the signal from the coin's surface takes part in the image formation, while the right image is produced by imaging the signal from the coin's back side that is modulated by the surface signal. Because there are no conspicuous differences between the surface and the back side signal, there seem to be no inclusions or cracks inside the coin. Measurement frequency is 25 MHz; image width is 20 mm.

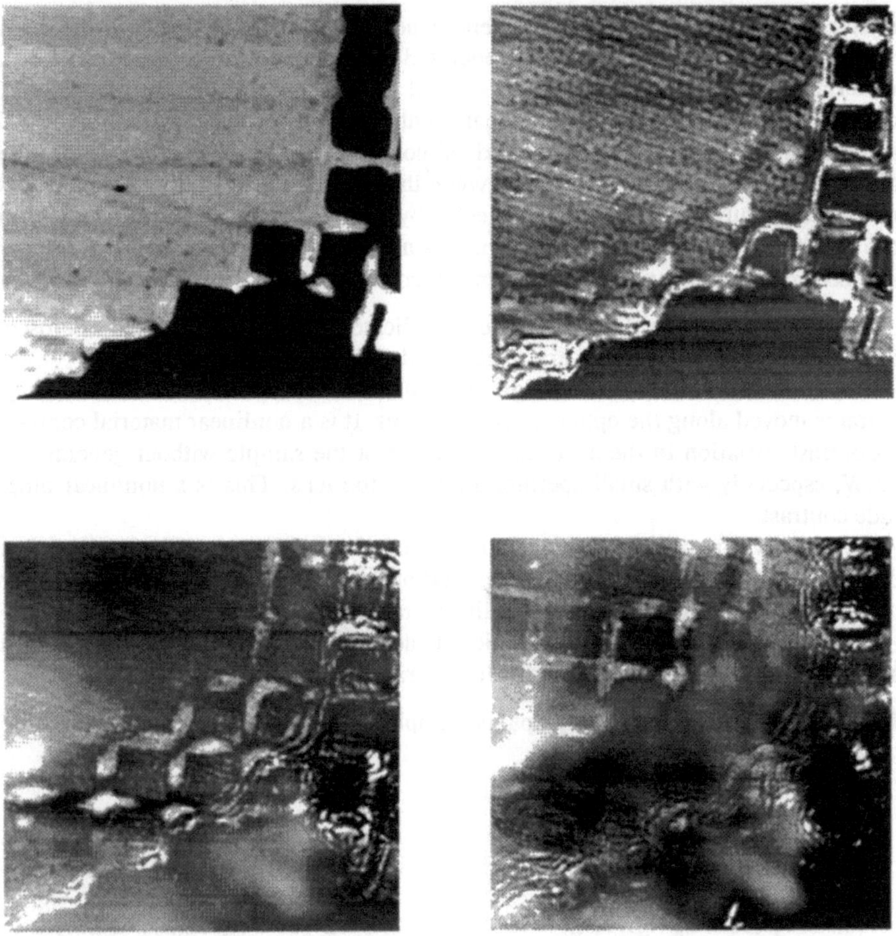

Fig. 2. Acoustic image of a copper lattice under a droplet of solder as an example for imaging depth structures. From the upper left to the lower right the defocus increases (i.e. the distance between acoustic lens and the sample decreases) and so lower structures of the sample can be detected. Measurement frequency is 200 MHz, image width is 1 mm.

References

1. Briggs A., *Acoustic Microscopy*, Clarendon Press, Oxford 1992.
2. Bereiter-Hahn J. *Akustische Mikroskopie*, in Robenek H. (Hrsg.) *Mikroskopie in Forschung und Praxis*, 261-323, GIT, Darmstadt 1995
3. Yu Z. and S. Boseck, Rev. Mod. Physics 67/4, 863-891 (1995)
4. Block H., S. Boseck, G. Heygster: Optik 86, 27-37, 1990.
5. Forgber E., G. Heygster, S. Boseck: Optik 87, 170-178, 1991.

Application of Tandem Scanning Microscopy to the Non-destructive Investigation of Tool Marks on Historic Ivory Carvings

Theo Jülich; Markus Miller; Lars Fiegenbaum

Hessisches Landesmuseum Darmstadt, Friedensplatz 1, D-64293 Darmstadt,
Tel.: (+49)6151-16-5785, Fax: (+49)6151-28942

Abstract. Development of a method for the dating of tool marks on the surface of works of arts. The surfaces of medieval sculptures and recent ivory-samples are scanned. Their digitised topographic information is analysed by statistical methods.

1 Investigation of works of art in ivory

The aim of this research-project at the Hessisches Landesmuseum in Darmstadt is the development of a method for the investigation and dating of tool-marks on the surface of works of art.

For the first step, the possibility of an examination of carvings in ivory by analysis of the surface information has been worked out. For dating reliefs and sculptures in elephant- and walrus-ivory or bone there does not yet exist satisfying techniques or methods.

In addition to the judgement of style and the chemical investigation of material, the analysis of tool-marks will become a justification for dating works of art consisting of bone or dental materials. For the art-historian almost all methods known to this day for dating carvings in ivory are unqualified. Especially concerning medieval objects, and some fakes of the nineteenth century; their classification is only by criteria of artistic style and sometimes leads to uncertain results. The analysis of raw materials - a procedure becoming more and more frequently used in historical sciences - may also may fail in this case, because it only can deliver the approximate date of the death of an elephant, walrus or cattle. In ancient treasuries raw ivory was freqently conserved over centuries, so that the material structure of an object may be much older than its artistic carving. In addition, those chemical analyses have to be declined because they generally require the taking of a material sample, inconceivable for example in the case of a small medieval object.

The investigation of tool marks by microscope attends to the process of the carving itself and its time period, and it represents a non-destructive way of testing. For these investigations a Tandem Scanning Light Microscope was used, which allows not only the magnification and scanning of the tool marks but also the recording of all topographic data for the computer.

Generally, the process of ivory-carving in historic times had not changed up until the invention of mechanical tools. The Roman craftsman made use of nearly the same kind of scrapers or gouges as the sculptor of the eighteenth century. Also the handling of the tools had not changed. Hardness and physical composition of ivory and bone demands a particular scraping manipulation of the tool. This continuity of the work-process guarantees similar and comparable tool-marks on medieval objects as well as on modern carvings: It is in any case the mark left by the cutting-edge of the tool passing the dental material which is often well conserved in hidden areas of the carvings.

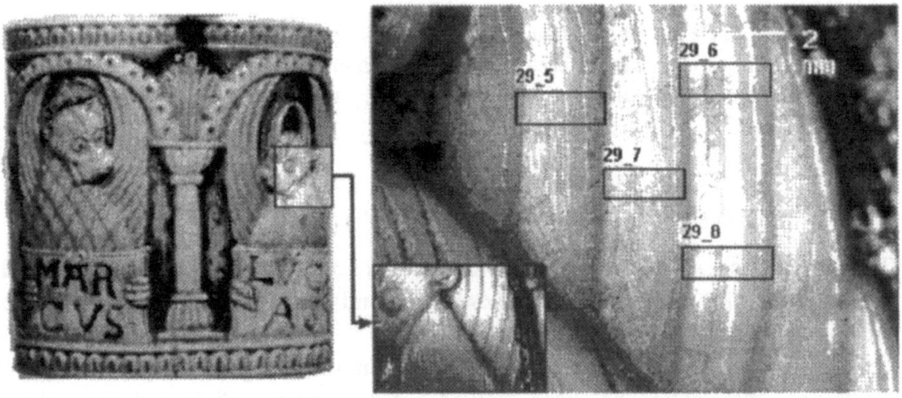

Fig. 1. Tool marks on a walrus-ivory box (Cologne, late 12th century)

As the cutting-edge is never perfectly smooth or straight, the jagged edge of the tool produces an uneven mark which is characteristic for one specific process of carving. Essential for the practice of the research work was the conjecture that those carving-marks show differences depending on the quality of the steel from which the tool is made. That means it had to be worked out to what extent tools with a less hard cutting edge leave different marks to those using a hard cutting steel or machines.
By evaluation of metal-analytical examinations of the archeo-metallurgists and by our own additional experiments it could be proved that iron tools before the 1500s were generally much softer than in later centuries. Considering the economical attitude of modern labour, one would assume that the fakes of the nineteenth century were produced with available high quality steel. The recognition of marks of 'less hard cutting steel '-tools therefore is a reason to believe that the carving was executed during the Middle Ages.
To find out whether the appearance of the tool mark is really dependent upon the quality of steel used for the cutting edge, a great number of tool mark samples were scanned with the TSM. They were effected with different tools of four degrees of hardness: Two scrapers were made from iron similar to a medieval tool in the degree of hardness; the other scrapers were made from a steel of a quality which is current in modern times.

Those marks produced under known experimental conditions now are compared with original medieval and modern scraper marks found on objects in the collection of the Hessisches Landesmuseum Darmstadt.

2 Scanning of toolmarks

For scanning one tool mark, the following process is used: Using a light microscope, magnification 10 to 80, as well as a video camera, scraper marks on the original surfaces extensive enough for image acquiring can be located. A video-print of this area allows the TSM to get in position and on this picture the areas in question can be marked to be found again. With the aid of the TSM, five to seven overlapping areas are acquired from one selected tool mark, so that the complete scraper mark is scanned widthways (this means at right angles to the movement of the tool).
The produced sequences of topography images are combined to a topography array. The array contains up to 1.5 million height values and the reduction of information is necessary for further processing.

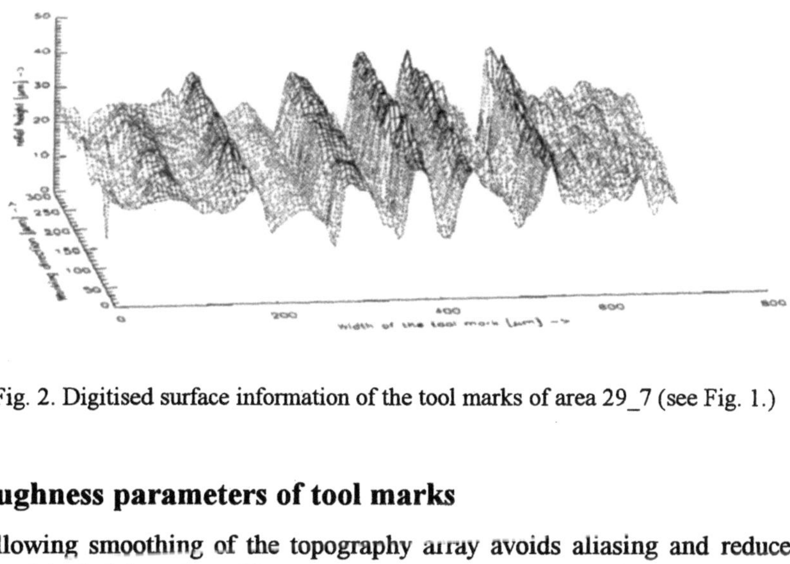

Fig. 2. Digitised surface information of the tool marks of area 29_7 (see Fig. 1.)

3 Roughness parameters of tool marks

The following smoothing of the topography array avoids aliasing and reduces the number of the height values. Then, based on extracted cross section profiles, parameters (e. g. roughness parameter) are calculated. Each parameter represents a certain feature of a tool mark. Subsequently, suitable ones are picked out to be used for the classification process. That leads to surface parameters, which allow the division of tool mark groups into 'hard'cutting steel, 'less hard' steel, 'machined', '19th century fakes' and 'medieval' tool marks. There are no examples for computer aided surface analysis of objets d'art, which switch over the material analysis. For this reason some surface parameters from mechanical engineering are used.
The considered roughness parameters are the mean roughness (R), mean slope (Δ), and mean wavelength (λ). The means are determined in the form of a mean absolute

deviation (Ra, Δa, λa) as well as in the form of a root mean square deviation (Rq, Δq, λq). Furthermore, the surface bearing index (Sbi), core fluid retention index (Sci), valley fluid retention index (Svi), the skew of the profile (Sk) and the kurtosis of the profile (K) are estimated.
The most functional parameters were Ra , Sci and Δa. With their aid a major part of the medieval tool marks could be separated from recent ones. The parameters Rq, Δq, λa, and λq did not deliver any further information.

4 Classification and dating of tool marks

As mentioned above, the computation of useful parameters is only an interim step towards assigning tool marks to groups. The concluding step is the classification of the combination of the parameters. The plotted values of the most functional parameters cluster according to the tool groups. In the classification process these clusters are used to determine the computational distance between the investigated tool mark and the tool mark groups.

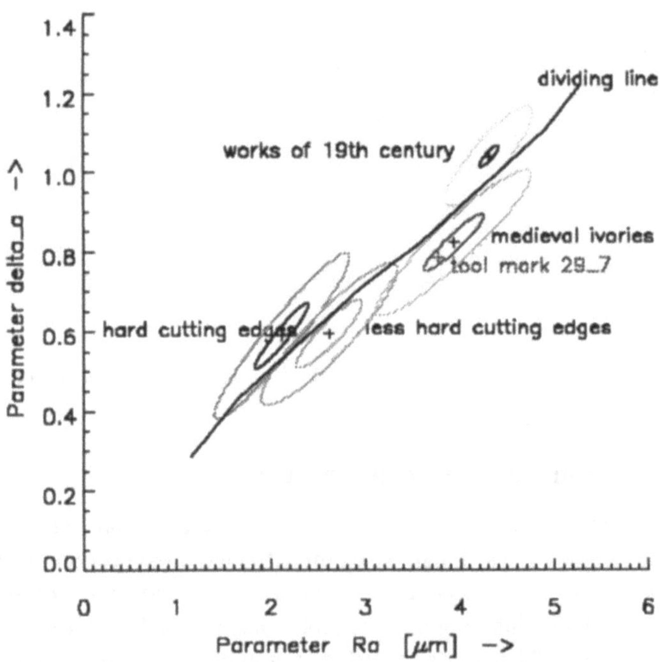

Fig. 3. Classification of the tool marks according to the quality of steel

The result of the classification allows one to conclude whether the investigated tool mark has been manufactured either by a hard or by a less hard cutting edge, and furthermore whether that the carving was executed either in recent times or during the Middle Ages.

Spectroscopical Techniques (Fluorescence, Luminescence, Reflection, Absorption, etc.)

Ground-Penetrating Radar Applications in Cultural Heritage

V. Pérez[1], F. García[2], J. Clapés[1], R. Osorio[1], J.O. Caselles[1], J.A. Canas[1] and Ll. Pujades[1]

[1] Servei de Geofísica Aplicada, C/ Gran Capitán s/n. Ed. D2. Univ. Politècnica de Catalunya, 08034 Barcelona. Spain.
[2] E.T.S.I. Geodesia, Cartografía y Topografía, Univ. Politècnica de Valencia, Valencia, Spain.

Abstract. Ground Penetrating Radar techniques were used in ancient churches to determine the vertical and horizontal position of possible pathologies in the walls or floors, and also the possible existence of other structures underneath the churches.

1 Introduction

The Applied Geophysical Service of the Politechnical University of Catalonia has carried out ground penetrating radar investigations at cultural heritage sites in Valencia (Spain). Two interesting cases are the integrated study of two churches, where three different types of problems were analysed during the research job.

GPR investigations are usually applied to civil engineering problems such as tunneling prospecting (Westerdahl et al., 1992; Canas et al., 1996), structural pathologies studies (García et al., 1996), and geotechnical problems (Doolite and Asmussen, 1992; Saarenko et al., 1992; García et al., 1996).

Two problems were civil engineering ones, concerning the actual state of the structures and their possible pathologies. Observed in the floor and in the walls of the ancient churches were humidity damage, materials detachment and underground layer movements because of deformations due to water pressure (in clay layers) or to the poor consistency of alluvial formations (gravels and sands).

The third problem, one of archeological research, involved the search for the possible existence of other structures or older cultural layers underneath the monuments investigated.

Ground Penetrating Radar is a technique for geophysical prospecting which permits the determination of changes in the electromagnetic parameters of the studied medium (floor, subsurface, walls or other materials media) (Lorenzo, 1994). It is a non-destructive technique, which is of great importance in analysing cultural and art heritage objects.

Radar data are a sequence of reflections produced at the contacts between materials possessing a contrast in their electromagnetic parameters.

The interpretation of these results usually requires additional information or archeological and historical knowledge of the site. It is an interesting technique for use in conjunction with archaeological research or restoration jobs.

2 The GPR Prospecting

Ground Penetrating Radar is a non-destructive prospecting system based on the emission of electromagnetic waves and the transmission of these waves in the studied medium. The reflection of this energy in the interfaces where dielectric parameters of the medium change is detected by the antenna. It has been used for several years to detect underground features and objects because of its high resolution.

The propagation of a plane electromagnetic wave in a homogenous medium is defined by the equation:

$$\nabla^2 \mathbf{E} = (i\omega\mu\sigma - \omega^2\mu\epsilon)\mathbf{E} = \gamma^2 \tag{1}$$

$$\nabla^2 \mathbf{H} = (i\omega\mu\sigma - \omega^2\mu\epsilon)\mathbf{H} = \gamma^2 \tag{2}$$

where $\mathbf{E}$ is the electric field (V/m), $\mathbf{H}$ is the intensity of the magnetic field (A/m), ω is the wave frecuency and ϵ, μ and σ are the electromagnetic parameters of the medium: the dielectric constant of the material, the magnetic permeability and the conductivity.

If the propagation is along the z-direction, the equation is:

$$\frac{\partial^2 \mathbf{E}(z,t)}{\partial z^2} = \gamma^2 \mathbf{E}(z,t) \tag{3}$$

$$\frac{\partial^2 \mathbf{H}(z,t)}{\partial z^2} = \gamma^2 \mathbf{H}(z,t) \tag{4}$$

where $\gamma = \alpha + i\beta$ is the propagation factor, defined by the expression:

$$\gamma^2 = i\omega\mu\sigma - \omega^2\mu\epsilon \tag{5}$$

The solutions of the equations (3) and (4) are:

$$\mathbf{E}(z,t) = \mathbf{E}_0 e^{\pm\gamma z} e^{i\omega t}\mathbf{i} \tag{6}$$

$$\mathbf{H}(z,t) = \mathbf{H}_0 e^{\pm\gamma z} e^{i\omega t}\mathbf{j} \tag{7}$$

with $\mathbf{i}$ and $\mathbf{j}$ unitary vectors in the x-direction and the y-direction.

The real parts of equations (6) and (7) are the solutions of the physical problem. These solutions represent an harmonic movement with an attenuation coeficient.

The velocity, v of this harmonic wave is

$$v = \frac{\omega}{\beta} = \frac{1}{\sqrt{\frac{\mu\epsilon}{2}\left(\sqrt{1 + \left(\frac{\sigma}{\omega\epsilon}\right)^2} + 1\right)}} \tag{8}$$

And the attenuation parameter:

$$\alpha = \omega \sqrt{\frac{\mu\epsilon}{2}\left(\sqrt{1+\left(\frac{\sigma}{\omega\epsilon}\right)^2}-1\right)} \tag{9}$$

The attenuation parameter α and the wave velocity v are frequency dependent, and also depend on the electromagnetic parameters of the medium: magnetic permitivity μ, dielectric constant ϵ and conductivity σ.

When the material behaves as a dielectric media (when $\sigma << \omega\epsilon$) and the relative magnetic permeability $\mu = 1$, the imaginary component of ϵ is zero and it is possible to simplify these solutions. The pulse velocity can be obtained for the expression:

$$v = \frac{c}{\sqrt{\epsilon_r}} \tag{10}$$

This approximation is valid for a range of frequencies between 10 MHz and 1 GHz (the ground penetrating radar range frequencies) but when the conductivity of the material is less than 10ms/m (Davis and Annan, 1989).

The contrast between electromagnetic parameters in different materials in the medium where the radar pulse is traveling produces reflections and refractions of the energy. The reflection and the transmission coefficients are the parameters which determine the amount of energy that is reflected in the interface and the amount of energy that is refracted and travels toward the next electromagnetic properties contrast. High contrasts produce strong reflections, and the wave arrives at the antenna with high amplitude.

The vertical resolution depends on the pulse frequency, as well as on the wave penetration in the medium. Low frequencies produce a lower resolution record than high frequencies, but on the other hand allow a deeper penetration than the high frequency record. Experimental measurements of attenuation versus frequency for several rock and sand types (Turner and Siggins, 1994) show that the energy attenuation is proportional to frequency and to electromagnetic parameters. In the same material, the wave attenuation for high frequencies is higher than the attenuation for low frequencies. The wave attenuation in materials with a high conductivity coefficient is larger than the attenuation in dielectric materials.

3 GPR research in San Jorge Church (Paiporta, Valencia)

San Jorge Church was built at the end of the XVIII century atop the remains of a more ancient church, probably of the XV century, which was demolished. It is of neoclassical style with some elements of poor Valencian baroque. In the IX century Paiporta was an arabian farmhouse. It is probable that the tower of the church was built over an ancient minaret. Now it is a village near Valencia City and San Jorge is its main church.

The possible restoration of the church and the problems with the humidity were the main reasons to study it with Ground Penetrating Radar in order

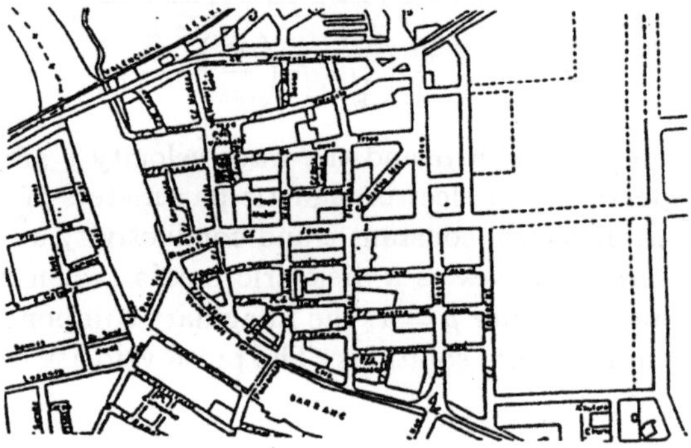

Fig. 1. Map of Paiporta (Valencia. Spain). The studied church is placed over alluvial materials near the Xiva river. San Jorge Church is marked with a circle.

to determine the damage areas caused by humidity and to study the possible existence of the previous church remains.

San Jorge Church is built on alluvial materials near the river. A visual inspection allows one to observe humidity damages in some areas of the floor and on the church's eastern wall. Also, it is easily possible to see detachment problems in the walls of the tower.

Additional information was obtained from stratigraphic columns of four boreholes placed near the site and by historical knowledge: ancient construction on the site and the existence of an ancient canal, today buried. This canal irrigated the Valencia fertile area.

The first stage of the GPR research was a subsurface study. It is important to detect stratigraphic anomalies that could damage the church structure because of deformations or movements.

To acquire the best resolution into the desired depth range, we selected a 100 MHz center frequency antenna and a 500 MHz center frequency antenna. Both were monostatic antennae. The first antenna can detect reflectors at 15 meters depth in the materials of this site. The second antenna can detect reflectors at three meters depth in the same materials. We completed five profiles in the floor. Every profile was done with the two antennas, yielding two record sections for each profile.

During this stage we observed a clear reflector situated about 1.60 meters deep (figure 3) caused by the contact between overlying materials and alluvial materials (gravel stratum).

It is important to note the existence of clay bodies (figure 4) in this gravel layer. This level occurs at 4.5 meters depth where a change in the radar record shows the contact with a clay and gravel mixing layer. About 9 meters deep, other reflectors indicate the existence of a very dispersive medium, probably a very plastic clay layer (figures 4 and 5).

The principal characteristic of this deeper layer is its deformation. In the radar records we can see that the contact is found about 12 meters deep near the main walls of the church and between 9 and 8 meters deep under the main nave of the church (figures 4 and 5). This deformation is probably caused by

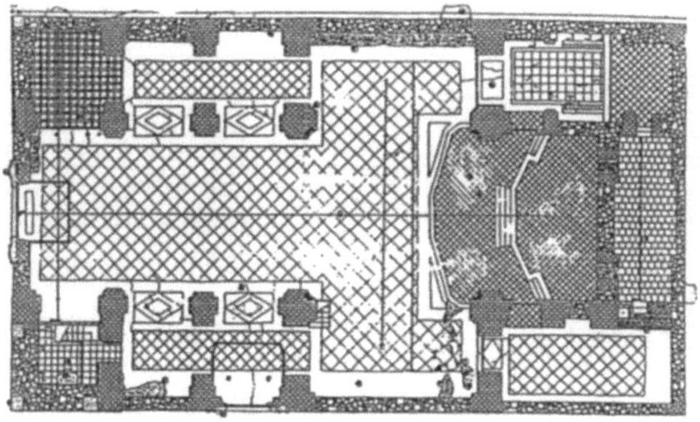

Fig. 2. Ground plan of San Jorge Church. The profiles to study the subsurface and to find cultural levels underneath the church are indicated over the floor

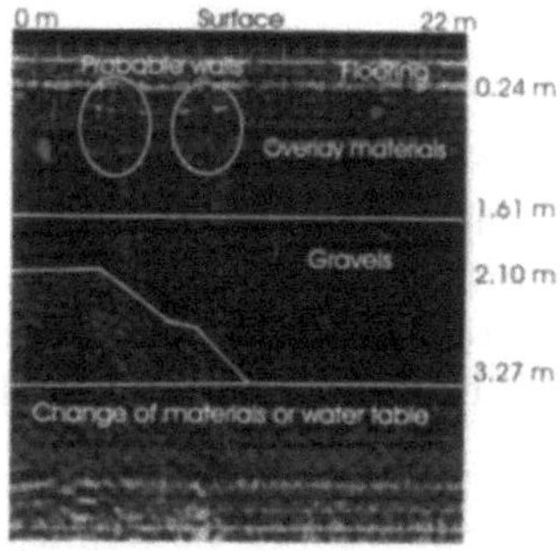

Fig. 3. The record with a 500 MHz center frequency antenna acquired along the main nave of the church. Four meters from the main door is the first reflection in the possible wall. This reflector is 2.5 meters wide. Nine meters from the door we notice the existence of other, similar reflectors. These begin at a depth of 0.25 meters in the fill. Over these reflectors, a change in the electromagnetic parameters in the medium is seen. A reflection produced at the contact between filling materials and gravel is recorded about 1.6 meters deep. Other reflector materials in the gravel layer are detected along the first twelve meters of the profile. Perhaps there are possible remains of the ancient church or of other cultural layers.

the weight of the walls. The water contained in the layer is probably a very important factor in this deformation. The process of this layer deformation can cause damages in the church structure such as voids or detachment problems.

In figure 4, the profile is the same as the profile in figure 1. In the same area where figure 1 shows the possible walls we can see strong reflectors (in white) which are probably the contacts between the floor and the possible walls. Figure 4 shows the existence of five possible layers. The top one is the church pavement. Under it there is a very reflective layer formed with fill materials. The third layer corresponds to gravels. The lowest section of this layer represents the clay and gravel mix. A reflection at a 4.5 meters depth shows a possible clay area placed in the gravels layer. The contact between the clay and gravel mix with very plastic clays is situated 10.5 meters deep in the start of the profile and 9 meters

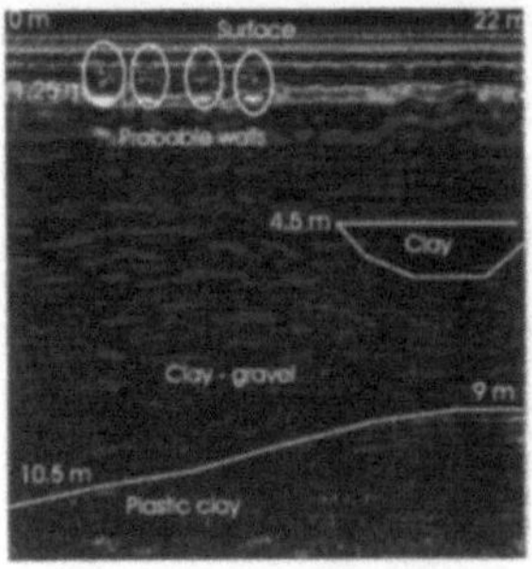

Fig. 4. The record at a 100 MHz center frequency antenna acquired along the main nave of the church.

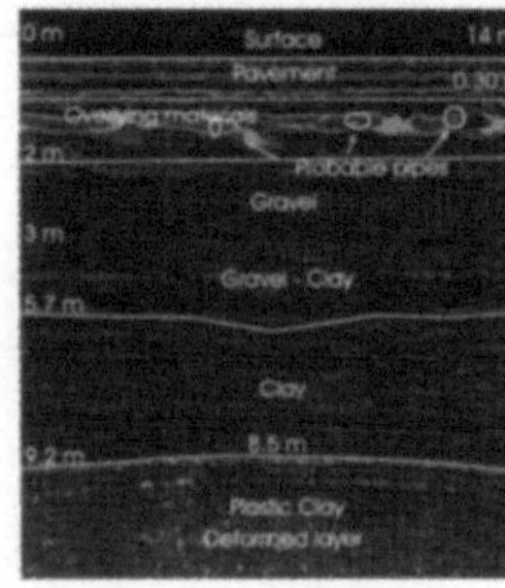

Fig. 5. Radar data obtained with the 100 MHz center frequency antenna in the profile across the main nave.

deep at the end of the profile.

In figure 5 the deeper reflector is a plastic clay layer. The other reflectors observed in this record are the contact between the pavement of the church and the overlain stratum, the contact between these materials and a gravel level, the gravel level and a gravel and clay mix and two clay layers.

The second effort was the study of the walls.

First, we studied the eastern wall. A visual inspection denotes damp areas on the surface (figure 6). To investigate inside the wall we selected a 900 MHz center frequency antenna. The radar data shows a wall affected by humidity in the lower part. The outside of the wall was the most affected area (figure 6).

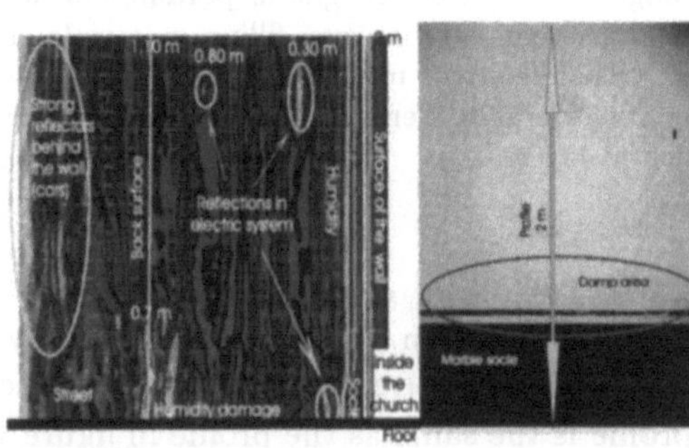

Fig. 6. The photography shows the eastern wall and the data obtained there. It is possible to observe the signature of the damp area in the lower part of the wall. The outside area is more affected by humidity than the inside area. Other reflections are caused by the electric system and by the marble inside the church in the lower part of the wall. Outside the wall there is also a strong reflector, caused by metallic bodies, probably the cars parked near the church.

The existence of an ancient canal, today buried, is well known. The placement of this canal could be along the street near the eastern wall. The search for the canal placement was done with a 500 MHz center frequency antenna. It was found bordering the eastern wall of the San Jorge Church. The radar data show a strong reflector which denotes a damp area along the wall (figure 7). There were two places where the humidity was greater. In these areas, the damage caused by the humidity was important, although the inside of the wall was not affected.

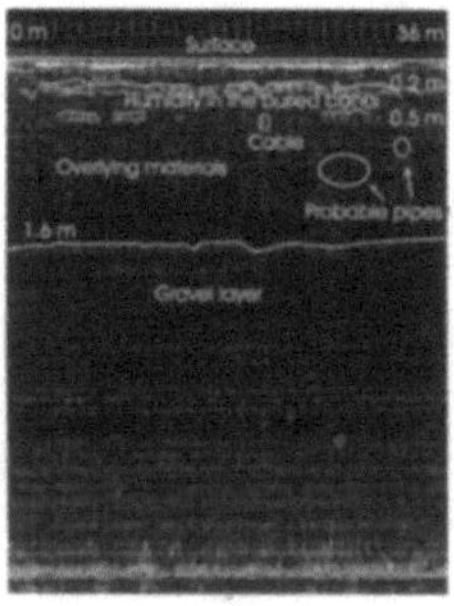

Fig. 7. Radar data obtained outside the church. The profile was acquired along the street adjacent to the eastern wall of the church. The buried canal placement was found here. This canal is probably what caused the moisture problems in this area of the church.

The next areas studied were the tower walls. Visual inspection denotes superficial detachment problems. The radar data display irregular and strong reflections caused by voids and detachment damage (figure 8).

Fig. 8. Radar data obtained in a wall of the tower. There are some reflections caused by detachment problems and possible voids within the half first meter.

These reflections were located from the surface to 0.5 meters deep inside the wall. They could be caused by slow and small subsurface deformations (figure 5).

The third part of the study was one of archaeological research. GPR was used in the archaeological investigations to study cavities and the bottom of important buildings (Kong et al., 1992). The study of the possible existence of cultural layers underneath the church is the initial step before beginning restoration jobs in the monument. Ground penetrating radar recorded that these structures had strong and characteristic reflectors without damage to the church. Historical and archaeological information about the studied area are necessary to complement GPR data.

During this stage, strong reflectors were recorded near the main door (in the northern wall of the church) adjacent to the square. The reflectors could be possible walls underneath the church (figure 3) and also a possible vault (figure 9), probably from an earlier church, which was located in the actual square and in the northern area of San Jorge Church. The ruins are located between one and four meters deep. They are in the alluvial gravel layer and in the overlying formations.

In addition to these reflectors, a strong reflector was recorded as the contact between the actual church pavement and the overlay materials.

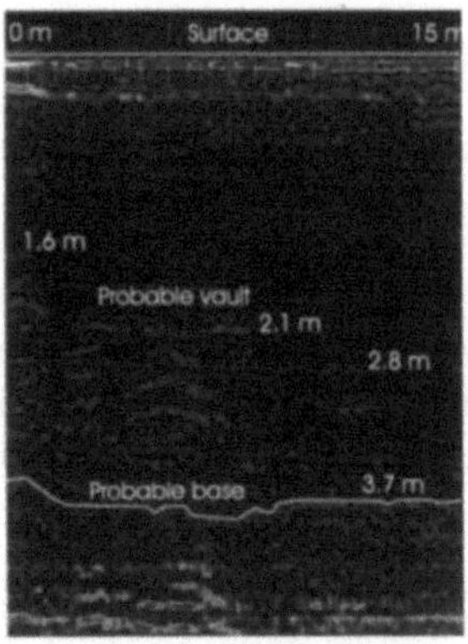

Fig. 9. Record acquired with a 500 MHz center frequency antenna, crossing the main nave near the door (one meter from it). A reflection like a vault is registered at about the third meter at a 1.6 meter depth. This is probably the remains of the oldest church, placed in the gravel layer.

4 GPR applications to the Temple (Valencia)

The temple, consisting of building and church, occupies the major area of an ancient monastery house of the Templars in Valencia. From the ancient building exists nothing but the name; so named because Jaime I conceded the tower of Ali Bufat and some nearby houses in 1240 to the Guillem templar of Cardona, next to the door of Bab-al-Sakhar, the place where the templars headquarters of Valencia, until the termination of this military order in 1313. A little later, after Jaime II created the Order of Montesa and transferred the properties of its extinguished Order to this order, the old Temple was the grand masters residence. This is the normally used name: "Palace of the Temple". It is not really correct. The correct name is the Sacred Royal Convent of Our Lady of Montesa. From 1587, the date that the grand master Pedro Luis Garceran of Borja gave up to their suzerainty, this was transferred by Sixto IV to the Crown, and the general lieutenants replaced the grand masters as residents in the statement palace. Soon after the earthquake of 1748 that destroyed the main castle-monastery of Montesa, king Fernando VI ordered the erection of a new and monumental building in the ancient temple that would also inherit the precedence among the Montesa monasteries. The direction and execution of the works (finished in times of Carlos III) were entrusted to the architect from the Court of Madrid, Miguel Fernandez, author therefore of one of the most characteristic buildings of the neoclassical style in Valencia. One is the distinguished Palace. The other is the Church. These consist in a great facade with two tower-steeples, frontal with allegorical statues of "The Religion" and "The Devotion" by the sculptor Jose Puchol, and a great royal shield on the piazza, all in very classical lines and according to the peculiar conception of the call "giant order". The dome and the tower, in concession to the ornamental tradition of the country, are covered with the usual tiles glazed in blue. The interior of the temple, wide and clear, configures a Latin cross plan: the greater altar standing with multicoloured jaspers, Antonio Gilabert's work; the chapel of San Jorge, the great stylistic unit; and the grand masters' Francisco Llausol of Romani and Bernardo Despuig sepulchral statues, in Renaissance style and carried from Montesa. The fresco paintings in the vault and apse were painted by Jose Vergara.

The geophysical investigations achieved by GPR techniques are framed in the works of the architectural evaluation that are being carried out in this temple of Our Lady of Montesa (in the well-known church of the temple) in the city of Valencia (Spain). The principal objective of this study is the location of possible anomalies due to underground ancient human constructions. These can affect the structure of the current building, which produces the apparition of fissures in walls and roof, with the associated problem of moisture that affects the different building areas and the fresco paintings.

The church is located in the perimeter area of the ancient nucleus of the city of Valencia. The mensurations were carried out along 15 profiles (figure 10) in two sectors of the church, in June of 1995. Geotechnical soundings, existing in bordering zones, were also utilized in order to determine the dielectrical constants of the present materials. This facilitates the calculation of the electromagnetics wave propagation velocity, so that we could determine the depth of the anomalies with greater exactness.

The utilized equipment was the system GPR SIR10, GSSI, with two screened transductor-antennae with 500 MHz and 100 MHz central frequencies.

Subsequently, the obtained records were processed in laboratory by the pro-

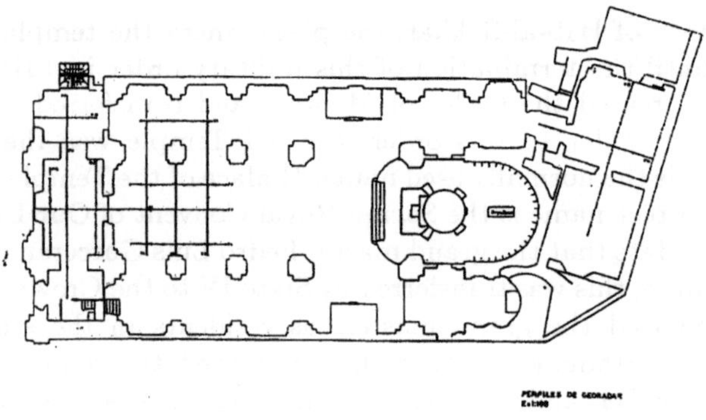

Fig. 10. Ground plan of the temple church. The 15 profiles realised during the study are marked on the plan.

gram RADAN III, applying the techniques of filtering in order to increment the ratio signal/noise and so to define the existent underground reflections with precision.

In this article, a brief series of radar data has been chosen. They present diverse interesting aspects of this study.

In figure 11 a strong singular reflection was recorded. It is owed to a metallic object. In zoom (figure 12), its geometry is better appreciated, and indicates the possible presence of an ancient tomb (the first grand masters were buried in coffins of lead), or an object with a bell form.

Fig. 11. Radar data acquired with a 500 MHz center frequency antenna in the west area (profile 11 in figure 10). A strong reflection was recorded, probably produced in a metallic object near the surface. It was produced between 1 and 1.6 meters depth.

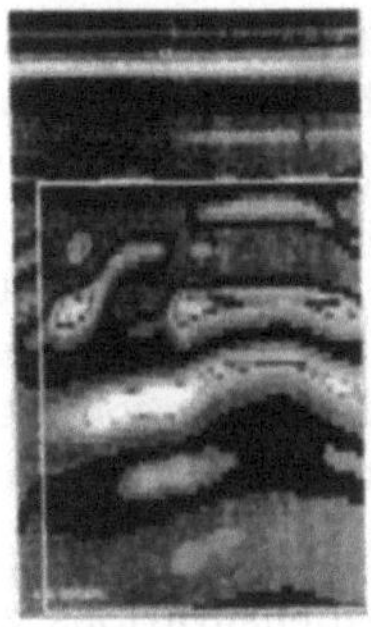

Fig. 12. Detail of figure 11. The geometry of the metallic object is best appreciated. It could indicate the presence of an ancient tomb or a bell form object.

In profile 15 (figure 13) the location of the ancient well that gave water to the complex of the Temple was recorded. Now it is buried and contains water in the bottom.

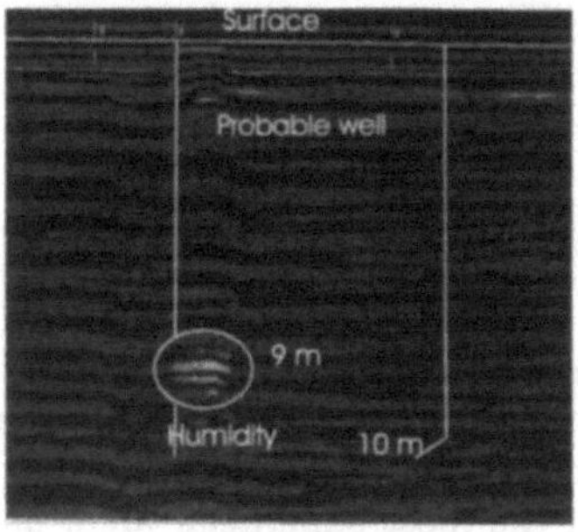

Fig. 13. Record data obtained in the east area of the church. The irregular subsurface and the strong reflection caused at 9 meters depth could be produced by an ancient well, today buried. This ancient well supplied water to the monastery of Montesa and before this study its location was not known. The strong reflection at 9 meters depth is probably because of humidity in the bottom of the buried well.

In the radar data obtained in all the profiles, we can observe several sequences of cultural strata that indicate it is possible to appreciate different historical epochs that have happened in the city of Valencia. A good example was the record obtained in the profile 2, in the west area of the church (figure 14).

During the study, ancient structures underneath the church were found. One of these structures was recorded in profile 3 in the western area of the church (figure 15). In this radar data, the location of a possible crypt is observed, with a series of steps in order to access it.

The applications of GPR method have not only contributed with valuable information to geotechnical problematic and architectural evaluation of the building for its restoration and conservation, but have also revealed great interesting archeological data for the cultural heritage of this emblematic building for the city of Valencia.

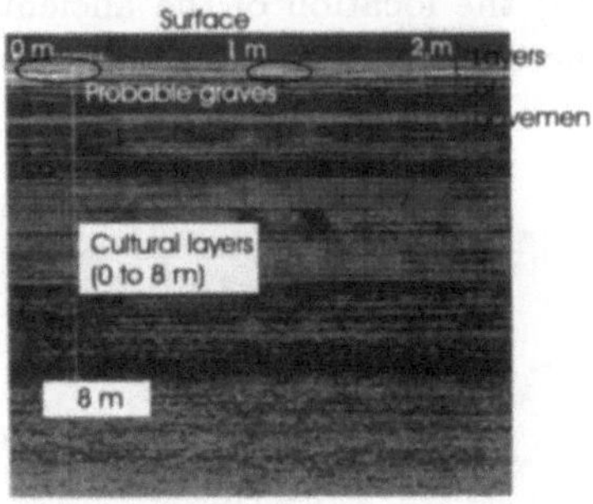

Fig. 14. Radar data obtained in profile 2. It is possible to appreciate the several sequencies of cultural strata that corrrespond to different historical epochs of the city: roman, middle age and modern epoch. In this record it is also possible to observe the different pavement layers in the most superficial area, and the strong reflections due to the tombs in the subsurface.

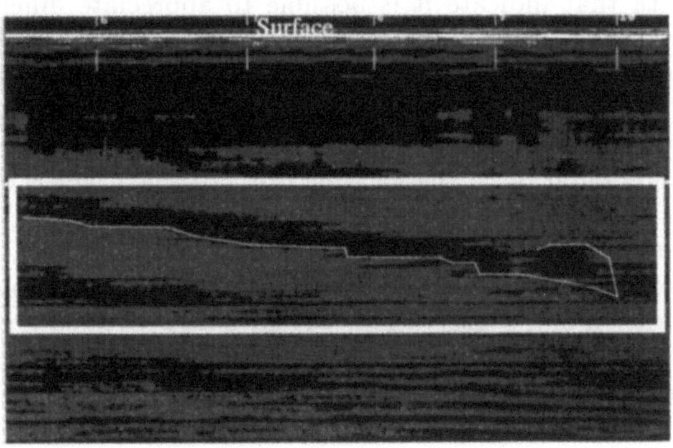

Fig. 15. Possible archeological remains between 2 meters and 8 meters depth. Radar data are a possible crypt and steps.

5 Conclusions

The GPR study of these two churches in Valencia detected moisture and detachment damage in the walls and the possible causes: layer deformation or movement and existence of humidity in a buried canal. Geotechnical studies with GPR in the temple church found the existence of different cultural levels in the subsurface, which are eight meters deep. An accurate study of the floor of the churches shows the existence of different pavements superimposed.

GPR provides high resolution for archeological research, but the cooperation with archeologists and historians is necessary to interpret the radar data as cultural remains.

A historical review is an important task prior to restoration work in cultural monuments. This can specify the oldest ruins beneath the structure. The review can also focus on probable causes of the different pathologies affecting the structure. Such a review also assists in the interpretation of the geophysical data as well as the design of the data acquisition.

Aknowledgements

The authors would like to thank all those who helped collecting radar data. A specific thanks to Cristina Ballester, Carlos Barberà, Luis de la Cruz and Fernando Benavent for providing the cartographic information and the maps, to Vicente Santamaría García for providing historical data and graphic information and to Dr. Hermann for his corrections.

References

1. Canas, J.A., García, F., Clapés, J., Osorio, R., Pérez, V., Pujades, L.G., Caselles, O., Ugalde, A.: Aplicaciones de geo-radar para la determinación del volumen de hormigón necesario para la reparación del revestimiento de la bóveda y los hastiales de un tunel. Geogaceta. **20(4)** (1996) 974–977
2. Davis, J.L. and Annan, A.P.:Ground-penetrating radar for high-resolution mapping of soil and rock stratigraphy. Geophysical Prospecting. **37** (1989) 531–551
3. Doolite, J.A. and Asmussen, L.: Ten years of applications of ground penetrating radar by the united states department of agriculture. Fourth International Conference on Ground Penetrating Radar. Finland. **16** (1992) 139–147
4. García, F., Canas, J.A., Clapés, J., Osorio, R., Pujades, L.G., Pérez, V. and Caselles, O.:Aplicaciones de geo-radar para determinar oquedades en el futuro emplazamiento de las pilas de sostenimiento de un viaducto. Geogaceta. **20(6)** (1996) 1361–1364
5. García, F., Canas, J.A., Clapés, J., Osorio, R., Pujades, L.G., Pérez, V., Caselles, O.:Aplicación de técnicas de geo-radar para el estudio del tapón de jet-grouting inyectado en los túneles del metro. Geogaceta.**20(6)** (1996) 1365–1368
6. Kong, F-N., Kristiansen, J. and By, T.L.: A radar investigation of pyramids. Fourth International Conference on Ground Penetrating Radar. Finland. **16** (1992) 345–349
7. Lorenzo, E.:Prospección geofísica de alta resolución mediante geo-radar. Aplicación a obras civiles. Tesis Doctoral. Universidad Politécnica de Madrid. (1994)

8. Saarenko, T., Hietala, K. and Salmi, T.: GPR applications in geotechnical investigations of peat for road survey purposes.Fourth International Conference on Ground Penetrating Radar Finland.**16** (1992) 293–305
9. Turner, G. and Siggins, A.F.:Constant Q attenuation of subsurface radar pulses. Geophysics. **59(8)** (1994) 1192–1200
10. Westerdahl, H., Austvik, A. and Kong, F-N.: Geo-radar in tunneling - The Tunnel Radar. Fourth International Conference on Ground Penetrating Radar. Finland.**16** (1992) 41–45
11. H. Frieser. *Photographische Informationsaufzeichnung.* R. Oldenbourg Verlag München Wien, Focal Press London - New-York, 1975.
12. Vision International. *Imagine OrthoMax User's Guide Version 8.2*, 1994.
13. Landolf Mauelshagen Werner Marten and Rainer Pallaske. Digital orthoimagsystem for arcitecture presentation. *ISPRS Journal of Photogrammetry and Remote Sensing*, pages 16–22, 1994.

Raman Spectra Treatment with Signal Processing Techniques

J.M.Yúfera, A.Ramos, O.Latorre, M.J.Soneira, S.Ruiz-Moreno and A.Rey
Signal Theory and Communications Dpt., Universitat Politécnica de Catalunya.
Campus Nord UPC Edifici D5, C/ Sor Eulalia de Anzizu, s/n. 08080 Barcelona
Tel. (34)(3) 4016442, Fax (34)(3) 4016447, E-mail: linjgf@tsc.upc.es

Abstract. In this communication we present different techniques for the reduction of the fluorescence and noise that obscure the information contained in the Raman spectra of pictorial materials.

1 Introduction: Raman Spectroscopy and Noise

Raman spectroscopy is a technique that allows the identification of pictorial materials. The information needed to identify the analysed material is, basically, in the correct location of Raman bands but if the spectrum has not been acquired in ideal conditions, the noise will always be present as a perturbation which can mask them.

In this communication different methods for noise reduction in Raman spectroscopy are used, based on filtering in the Fourier domain of the studied spectrum and the application of the maximum entropy method.

Among the different kinds of noise that exist when the signal is detected, fluorescence, shot noise, thermal noise and cosmic noise stand out. Both shot and thermal noise are represented in a spectrum as a ripple which can come to hide the Raman bands or to be confused with one of them, making easier, in this way, a possible ambiguity in the identification of the analysed pictorial material. On the other hand, the cosmic noise will be observed, in the Raman spectra, as a high intensity peak (spikes) with a narrow bandwidth. It is difficult to confuse a cosmic ray with a Raman band because of its characteristics but the problem is that the methods for noise reduction that we present increase the bandwidth of the spikes creating possible false Raman bands. For all these reasons it is necessary to use a work tool that improves the quality of the Raman spectra obtained in the laboratory: the signal processing.

2 Signal Processing Techniques for Noise Reduction

Among the techniques used for fluorescence reduction, we can emphasize two [1]: the edge detection method and the digital filtering techniques. Even though each one of these techniques can be applied separately as a fluorescence reduction technique, the combined application of both of them allows us to obtain more accurate results.

First, we propose to reduce the shot noise of the spectrum with a low pass filter (Papoulis window), and to determine its optimum cut off frequency we use a Bush algorithm [2] modification. On the other hand, in order to know a first estimation of the position, width, and intensity of the Raman peaks of the studied spectrum, the edge detection method is used. With these parameters, and considering lorentzian peaks, an ideal Raman spectrum is generated, which is subtracted from the experimental one obtaining the fluorescence spectrum. By means of its FFT, its spectral width is estimated and therefore the cut off frequency of the high pass filter that will eliminate the fluorescence from the analysed spectrum is determined. Repeating the Marquardt algorithm on the filtered spectrum, a more precise Raman spectrum will be obtained, which will allow the identification of the pigment studied.

In addition, we have also applied the Maximum Entropy Method (MEM) [3] for ripple reduction. Each data treatment technique employed in the spectroscopy information recovery consists in finding the best estimation of the Raman spectrum from the measured spectrum and the information regarding the equipment and noise. The MEM gives a criteria for choosing a spectrum estimation among all the possibilities.

Finally, in order to eliminate the cosmic noise we use a simple algorithm which consists in shifting the contaminated spectrum and uses the slow variation of the Raman bands to locate the cosmic spikes and eliminate them by least square fitting.

3 Application Examples: Results and Discussion

For testing the fluorescence reduction technique we processed a spectrum containing ultramarine blue and cadmium yellow from a green sample. Figure 1a shows the original spectrum as it was obtained at the laboratory and a strong fluorescence band present in every optical frequency can be seen. Table 1 shows the theoretical position of the Raman peaks. In this spectrum the shot noise was filtered, after previously determining the cut off frequency of the Papoulis window (figure 1b).

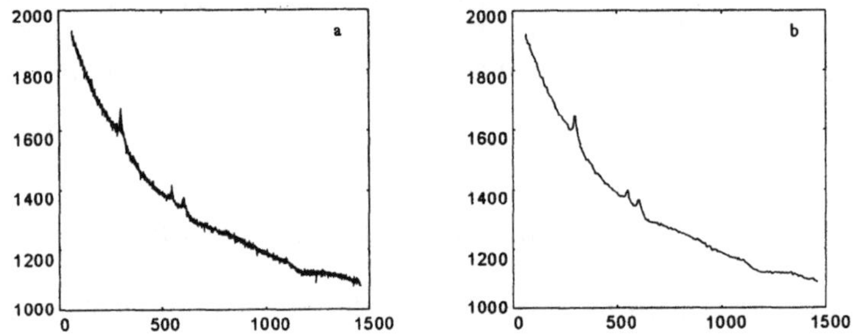

Fig. 1. a) Cadmium yellow and Ultramarine blue spectrum. b) Filtered spectrum.

Following the method explained above we can present, in table 1, the positions obtained with edge detection and the combination of edge detection and digital filtering. A better location can be seen.

On the other hand, figure 2 shows the Raman spectra obtained from chalk measured with 60 seconds of acquisition time and 4 scans (curve a), with 10 seconds and 3 scans (curve b), and that obtained, from this last one, after applying cosmic rays elimination and MEM (curve c). It can be observed how the location of the characteristic Raman bands of the material is easier after having applied this method.

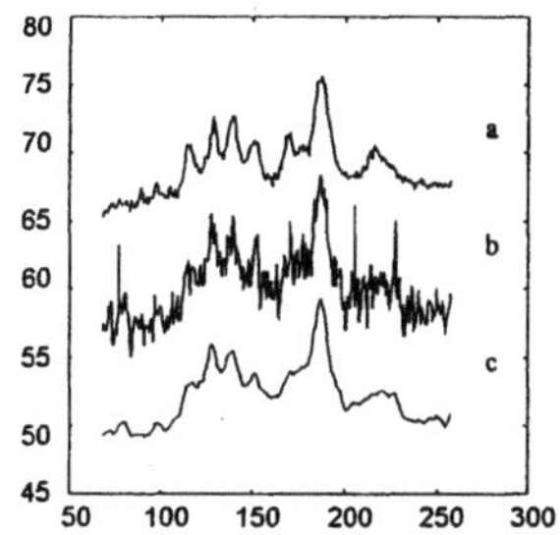

Fig. 2. Spectra from a chalk sample.

Table 1. Raman bands position depending on the applied technique.

	Theoretical	Edge Detection	Edge Detection & Digital Filtering
Ultramarine Blue	543	546.47	550.30
Cadmium Yellow	305	297.7	301.31
	609	600.42	604.58

4 Conclusions

The different signal processing techniques described in this communication have been demonstrated to be effective methods for fluorescence, cosmic and shot noise reduction in Raman spectra. They improve its signal to noise ratio, without having to use additional components, either optical or electronics, to those that a Raman system incorporates. The edge detection in combination with fast fourier transform filtering techniques shows that the results obtained are more accurate in the determination of the Raman peaks location than those obtained by both techniques applied independently. On the other hand, the application of maximum entropy method is compensated, clearly, through a lower time of sample exposition to the laser beam and, in some cases, allows the detection of the existence of a band apparently hidden.

Note that before applying any of the reduction techniques shown here it is necessary to apply previously a method to eliminate the cosmic rays in order to avoid the possible confusion of them with Raman bands.

References

1 P.A.Mosier-Boss, S.H.Lieberman, and R.Newbery, Appl. Spectr., **49**, p.630 (1995).

2 C.A.Bush, Anal. Chem., **46**, 7, p.890 (1974).

3 L.S. Greek, H.G. Shulze et al., App. Spectr., **49**, p.425 (1995).

Imaging Spectroscopy of Drawing and Painting Materials in the Near Infrared

F. Bayerer and A. Burmester
Doerner-Institut, Bayerische Staatsgemäldesammlungen
Barer Str.29, D-80799 München, Germany

Abstract. The MUßINI project developed a system for imaging spectroscopy on art objects. Besides improved spectral facilities to differentiate drawing and painting materials, a new quality of infrared reflectography has been achieved.

1 Introduction

A system suitable for obtaining spectral as well as spatial information on art objects was developed in the MUßINI project [1]. The system, working in the wavelength range of 300 - 1800 nm, includes an illumination monochromator, an infrared-vidicon, and respectively a Si-CCD camera for the lower wavelength range and a precise computer controlled 3d-positioning unit. The acquired data result in a 3-dimensional datacube with low wavelength, but high spatial resolution. The evaluation of the datacube is done by means of multivariate statistical methods, which have been especially adapted for the purposes of imaging spectroscopy on art objects. Using these facilities, many applications have been carried out.

2 Results

The applications can be divided into three main groups:

(1) The spectral facilities of the system help to differentiate materials which are indistinguishable for human observers. This ability can be used for the differentiation of drawing inks, minerals and some artist's pigments [2]. However, the broad band characteristics of the reflectance spectra in the UV, VIS and Near Infrared of most of the materials prevent an unambiguous identification which is often needed.

(2) In different wavelength ranges, the imaging component of the system allows an easy visualisation of the art object which is an important advantage for the interpretation of the observations. A typical application is the detection of an underdrawing hidden below non-transparent paint-layers, such as on a painting by Lucas Cranach. By moving the infrared-vidicon with the the 3-d positioning unit [3], 242 sub-images covering the whole area of this large painting (138 x 99 cm) have been acquired which results in a high spatial resolution of 10 lines per mm. The wavelength range is centered around

1650 nm with a bandwidth of about 50 nm. After acquisition, all sub-images were mosaiced together by complex software routines. The resulting infrared reflectography (Fig.1.), which reveals a carefully composed underdrawing, has recently been discussed from an art historian's point of view [4]. The system, as used for this application, is an extension and considerable improvement of conventional infrared-reflectography [5].

(3) Finally, combining the spectral and the imaging facilities, the system has been successfully applied to the investigation of drawings. A prominent example is the investigation of the 'Munich Rembrandt Apocrypha' whereby different drawing materials which were visually indistinguishable could be detected [6]. Again, the drawings have been imaged with high spatial resolution but additionally at 29 different wavelengths. The huge amount of data in the resulting data cube is efficiently compressed by factor analysis. The statistical treatment of the spectrally and spatially resolved data finally yields a factor image which shows the inks applied in a representation of different gray tones. The fact that the factor image resembles the original drawing facilitates the evaluation by the art historian.

These and other applications not reported due to limited space underline the multiple uses of the MUßINI imaging system in the humanities.

References

1 Funded by the Bundesminister für Forschung und Technologie, Bonn (No. 03-BU9MUE). The authors sign responsible for the content of this publication.

2 F. Bayerer and A. Burmester, *Nondestructive Imaging Spectroscopy of Drawings and Paintings,* Preprints of the 18th International Symposium on the Conservation and Restoration of Cultural Property - Spectrometric Examination in Conservation, Tokyo, 1994, p.76-86

3 The 3-d positioning unit is part of the VASARI system described in A. Burmester et al., *The Examination of Paintings by Digital Image Analysis,* 3rd International Conference on Non Destructive Testing, Microanalytical Methods and Environmental Evaluation for Study and Conservation of Works of Art, Viterbo, 1992, p.201-214

4 *Die Kreuzigung Christi von Lucas Cranach d.Ä. aus dem Jahre 1503,* Jahresbericht 1995 der Bayerischen Staatsgemäldesammlungen, p.11-22

5 A. Burmester and F. Bayerer, *Towards Improved Infrared-Reflectograms,* Studies in Conservation 38, 1993, p.145-154

6 A. Burmester and F. Bayerer, *Remote Imaging Spectroscopy of Drawings,* Proceedings of the 4th International Congress on Non-Destructive Testing of Artistic and Cultural Objects, Berlin, 1994, p.183-192,and earlier publications such as K. Renger and A. Burmester, *The Munich Rembrandt Forgeries Reconsidered: A New Technical Approach to the Investigation of Drawings*, Masterdrawings, 23/24 (4), 1985/86, p.526-537

Fig. 1. Infrared reflectography of Lucas Cranach d.Ä., *Christus am Kreuz*, 1503 (Bayerische Staatsgemäldesammlungen München, Inv.Nr. 1416)

TL Dating of Quartz Beyond 100 ka?

Peter Karelin

Forschungsstelle Archäometrie der Heidelberger Akademie der Wissenschaften, MPI Kernphysik, Postfach 10 39 80, 69029 Heidelberg, Germany

Abstract. Age determinations of sediments by thermoluminescence (TL) using the quartz fraction of sand can be reliable in the case of young samples. Using special detection filter combinations, the upper dating limit can be extended up to several hundred ka. Samples from the palaeolithic sites of Neumark–Nord and Schöningen (Germany) could be successfully dated up to nearly 400 ka.

1 Introduction to thermoluminescence

When ionizing radiation (α-, β-, γ-, cosmic) is incident on insulating crystals, some of the energy is stored by the lattice at crystal defects. Upon heating the crystals, this energy is released and a fraction of it is emitted as visible light, prior to the onset of black-body radiation: this is *thermoluminescence* (TL). The amount of light emitted is proportional to the radiation dose previously absorbed by the crystal.

An application of this phenomenon is the *age determination* of pottery or sediments [1]. The TL signal (or stored energy) is reset by heating (in the case of pottery) or by bleaching with sunlight (sediments). Then the accumulation of stored energy starts again and the absorbed dose can be determined in the laboratory. From the measured *absorbed dose* and the observed *dose rate* of the sample and its environment, the date of the last heating (pottery) resp. bleaching (sediment) is estimated (age = absorbed dose / dose rate).

The amount of energy that can be stored in the crystals is not endless, therefore there is a limitation to dating depending on the mineral and the dose rate. One can date the deposition of sediments reliably up to about 100 ka, but beyond that, problems arise due to saturation effects.

2 Measurements

The light is emitted from various lattice defects and, thus, can be observed in different glow peaks (see emission spectrum of NN3 Qz, fig. 1) having different properties (e.g. the wavelength of the emission, long term stability, bleaching behaviour). The aim of TL measurements is to separate the different peaks from one another in order to correctly determine the absorbed dose from a single peak with known properties, rather than from an overlap of several peaks with different properties. This is achieved by using certain optical filter combinations for the different mineral fractions. For quartz, the best combination found is a Schott MUG-2 plus a Hoya U340.

Equipment:

Elsec TL System 7188 with photomultiplier EMI 9635 QA
Detection filters: Schott MUG-2, Chance-Pilkington HA-3 (heat absorber)

Daybreak TL System 1150 with EMI 9635 B
Detection filters: Schott MUG-2, Hoya U340

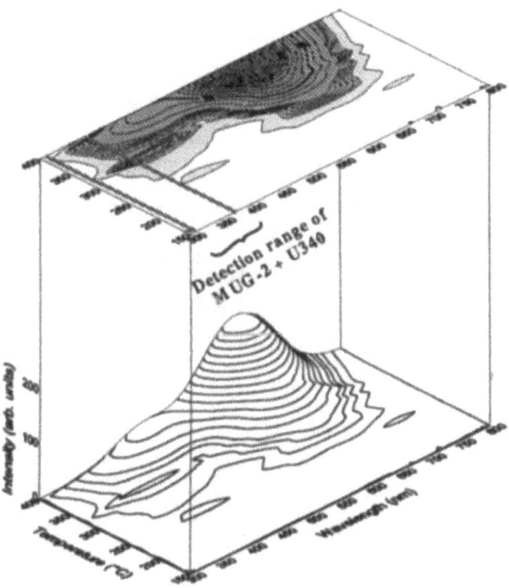

Figure 1. Emission spectrum of NN3 Qz.

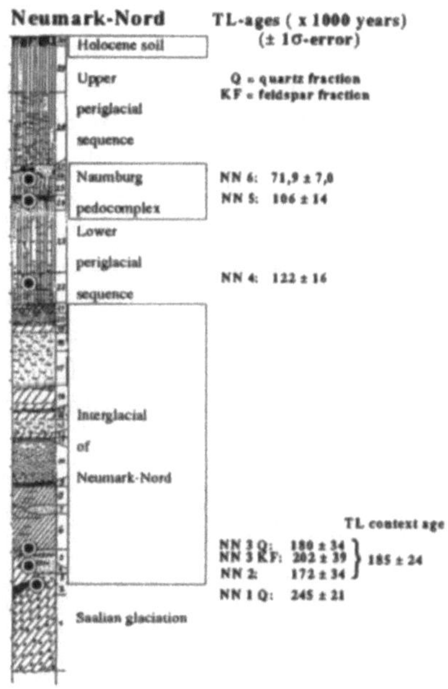

Figure 2. Neumark–Nord.

3 Applications

3.1 Neumark–Nord

The site of Neumark–Nord is an important area for the climatic stratigraphy of the last glaciations in Northern Germany. It is situated between the southernmost ice lobes of the Saalian and Weichselian glaciations.

Figure 2 shows the sediment structure and the sampling sites. The TL ages are in accordance with the stratigraphic classification of MANIA [3]: Weichselian loess situated above the fossil humic gley (NN6: 71.9 ± 7.0 ka) and Warthian basin loess below the gley (NN5: 106 ± 14 ka and NN4: 122 ± 16 ka). It cannot be definitely clarified if this humic gley represents the Eemian interglacial or a lower Weichselian interstadial.

The interglacial of Neumark–Nord overlies the Drenthe till. Both the fine grained basin facies (NN2: 172 ± 34 ka) and the coarse grained shore facies (NN3Qz: 180 ± 34 ka, NN3KF: 202 ± 39 ka) were dated as well as the Drenthian late glacial meltwater sands (NN1Qz: 245 ± 21 ka). This age is significantly higher than the age of the $\delta^{18}O$ stage 6 [2] (186–127 ka) even within the confidence level of 2-σ error (95%). The TL context age of NN2 and NN3 (137–233 ka) is higher than the age of $\delta^{18}O$ stage 5 (127–71 ka). This demonstrates a correlation of the interglacial of Neumark–Nord with $\delta^{18}O$ stage 7 (242–186 ka).

3.2 Schöningen

In this brown coal surface mining area of 350 000 m^2, very complete middle and upper Pleistocene profiles were found. They contain artefacts; from those made by early hominids (Holstein interglacial) up to artefacts from Roman times.

The sediments of the Reinsdorf and the Schöningen interglacials are situated *below* the Drenthe till. Thus they are older than the interglacial of Neumark–Nord which is situated *above* the Drenthe till (see above).

The ages from the silty mud (SCH12C2: 287 ± 52 ka (TL) and 256 ± 63 ka (IRSL)) overlying the archaeological layer and from underlying sands (SCH12C1 Qz: 385 ± 67 ka) imply that the Reinsdorf interglacial can be attributed to $\delta^{18}O$ stage 9 (301–334 ka). TL ages of sands superposing the Elsterian till (beginning of the Holstein interglacial) (SCH13 Qz: 378 ± 47 ka, SCH13 KF: 387 ± 92 ka) confirm that the Holstein interglacial is older (stage 11) than previously assumed.

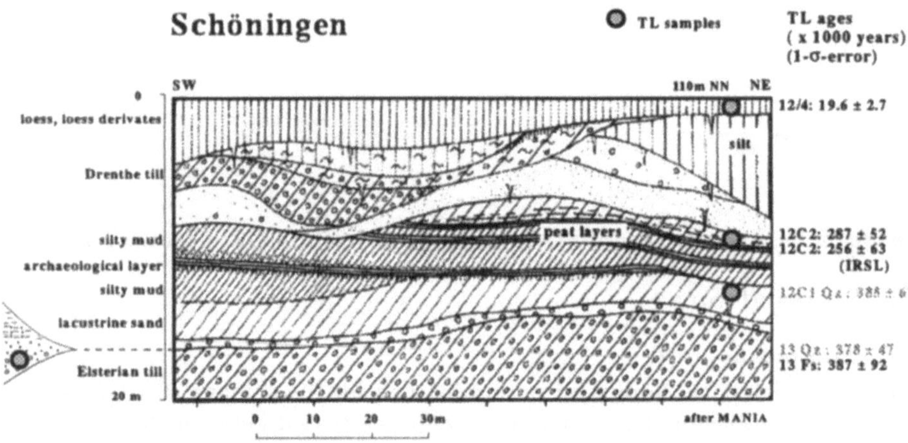

Figure 3. Schöningen: sediment structure and sampling sites.

References

1 Aitken, M.J., *Thermoluminescence Dating*, Academic Press, London, 359p, 1985.

2 Bassinot, F.C., Labeyrie, L.D., Vincent, E., Quidelleur, X., Shackleton, N.J. & Lancelot, Y., *The astronomical theory of climate and the age of the Brunhes–Matuyama magnetic reversal*, EPSL, **126**, 91–108, 1994.

3 Mania, D., *Stratigraphie, Ökologie und mittelpaläolithische Jagdbefunde des Interglazials von Neumark–Nord (Geiseltal)*, Veröff. Landesmus. Vorgesch. Halle, **43**, 9–130, 1990.

Separation and Identification of Raman Spectra for the Recognition of Pictorial Pigments

M.J. Manzaneda, S.Ruiz-Moreno, M. Breitman* and A. Tuldrá*

Signal Theory and Communications Dpt.; Applied Mathematics and Telematics Dpt.*
Universitat Politècnica de Catalunya. C/ Sor Eulalia de Anzizu s/n, 08034 Barcelona, Spain.
Tel. (3) 401 64 42 Fax (3) 401 64 47 E-mail: linjgf@tsc.upc.es

Abstract. In this contribution, factor analysis techniques have been applied to the separation of Raman spectra from mixtures of pigments. Finally, they are compared with those in a specific data base of standard pigments to identify the materials in the sample. Some real examples are shown.

1 Introduction

Raman spectroscopy, based on the Raman effect, is a versatile technique for the identification of pigments in an artwork with an excellent spatial resolution [1], [2], [3]. However, in most cases we cannot obtain the spectrum of a single pigment. Then, it is often of great help to separate the individual spectra from the spectra of the mixtures. The purposes of factor analysis are to obtain the number of components (or factors) in the sample, the spectra of these pure factors, and finally, the concentration of each material in every sample. Later, for an automatic identification, an algorithm compares the obtained spectra with those in a data base of standard pigments.

2 The Method of Factor Analysis

Due to the linearity of Raman spectroscopy, the spectrum of a mixture is a linear combination of the spectra of the components. However, from one single linear combination it is impossible to distinguish the contributions of one or more pure spectrums. Therefore, we need at least as many spectra of different samples (m) as pure components there are in every sample (n): $m \geq n$. Arranging the m spectra (of N points) of the samples in the columns of a matrix **D**, the spectra of the n pure components in the columns of a matrix **R**, and the concentrations of the samples in those of a matrix **C**, we can write the matrix expression:

$$\mathbf{D}\ [N \times m] = \mathbf{R}\ [N \times n]\ \mathbf{C}\ [n \times m]$$

Therefore, the problem converts into the calculation of the matrices **R** and **C** from the experimental matrix **D** [4], [5].

3 Experimental Results

In a simulation of the mixtures of spectra, that is, by summing mathematically the spectra of the pure components, the method works perfectly and is able to recover the original

Laser Photons in Medical Diagnosis

Theodore G. Papazoglou

Laser and Application Division
Institute of Electronic Stucture & Laser
Foundation for Research And Technology - Hellas
P.O. Box 1527, GR 711 10 Iraklion, Greece
E-mail: ted@iesl.forth.gr

1. Introduction

Various spectroscopic techniques have been investigated for optical diagnosis. Their target is the determination of information about the biochemical composition and/or the structure of the tissue. This approach is useful for the early detection of cancer as well as the diagnosis of other tissue abnormalities such as atherosclerosis, various infections etc. The motivation is the elimination of biopsy tissue samples while in situ evaluation of tissue is recorded in vivo by an optical fiber probe placed on the surface of the suspected tissue.

Fluorescence spectroscopy has been used with and without the aid of exogenous photosensitizers that target malignant tissue. The latter, however, is not an ideal solution for routine examination, since the administration of the exogenous photosensitizer could be followed by undesirable side effects. The use of intrinsic (usually UV or visible induced) tissue fluorescence may serve as a non invasive diagnostic technique. The results of several studies involving large sample tests demonstrate levels of reliability that range from unacceptable (<75% with many false positive predictions) to promising (>90% with few false negative detection runs).

2. Natural Chromophores Aided Spectroscopy

The main chromophores responsible for this tissue natural fluorescence signal are elastin, collagen, NADH and flavins[1,2]. Porphyrins reach peak emission at 675 and 610 respectively, if excited at 400 nm[3]. The major molecules in biological tissue which contribute to the fluorescence signal under 337 nm (pulsed nitrogen laser) near UV light excitation have been identified[4]. Tryptophan has an emission peak at ~390 nm, collagen at ~390 nm, chromophores in elastin at ~410 nm, nicotinamide adenine dinucleotide at ~470 nm, flavins at ~520 nm and melanine at ~540 nm[5-7]. Fu-Shih Pan reported in 1990 that the fluorescence intensity of NADH is directly proportional to the concentration of the fluorophore when excited at 337 nm (pulsed nitrogen laser), and that NADH had an emission peak at

480 nm[8]. Schwartz in 1974 suggested that the inability of cells to elevate their NAD^+ : NADH ratio at confluency is a characteristic of transformed cells and related to their defective growth control. The free NAD^+ : NADH ratio can be calculated and it is lower in hepatomas than in normal liver[9]. Surface fluorescence has been used to measure the relative level of NADH in both perfused and in vivo in tissues[10,11]. For blue excitation light, a strong natural fluorescence signal peaking at around 525 nm with a long tail reaching into the 700 nm region was observed for most tissues, aside from hemoglobin fluorescence. It has been shown that this green natural fluorescence signal is weaker in tumors than in normal tissue[5,12,13]. A possible explanation is differences in blood supply between metastatic and normal tissue. The spectral changes are dependent upon the grade of the neoplastic transformations[14,15,16]. Development of neoplastic tissue leads to increased light absorption and scattering, and a noisy and structureless spectrum where the contributions of the various tissue components are poorly defined. The differences of excitation spectra from normal and cancerous tissues is due to the filtering effect of other chromophores in the cancerous cell and tissues (Hb) as compared to the normal[14,15,5,3,12,17]. Richards-Kortum and her group developed a method for defining diagnostic algorithms for pathologic conditions based on fluorescence spectroscopy. This method uses fluorescencence excitation-emission matrices (EEM) to identify optimal excitation regions for obtaining fluorescence emission spectra which can be used to differentiate normal and pathological tissue. It is desirable to detect microscopic and biochemical changes of pre-cancer in order to identify patients at risk for developing invasive carcinoma. Currently available clinical techniques rely on global metabolic measurements of oxygen supply and delivery. They are therefore unsuited for early detection and continuous monitoring of local deficiencies in malignant tissue at the level of an organ or a tissue bed. Such desirable features could be attained through optical techniques.

3. Photosensitiser Aided Spectroscopy

Photoactivated, tumor marking agents, such as hematoprophyrin derivative (HpD), exhibit characteristic fluorescence spectra that are easily distinguished from natural fluorescence of the tissue. An ideal photosensitizing tumor-seeking agent should be pure, stable, non-toxic, have no cutaneous photosensitivity, and have a strong absorption and large fluorescence quantum yield in the far visible or near infrared ranges, using available light sources . Fluorescence detection as a diagnostic tool is based on the principle that HpD, when excited in the Soret band of violet light at 405 nm, emits in the red spectral region with a characteristic signature at 630 and 690 nm[18]. However, because excitation at 400 nm also leads to fluorescence from other chromophores (e.g. hemoglobin), various methods, such as background subtraction and normalization for autofluorescence (ratio-fluorometry), have been developed to eliminate these potential interferences[19,20]. The agent most widely used in clinical studies of photodynamic detection (PDD) is a purified mixture of dihemetoporphyrin ether/ester, and DHE, commercially known as Photofrin II

(QLT)[21]. Although it is not clearly understood why the DHE is selectively retained in tumors compared to normal tissue, a large number of studies have reported the use of fluorescence excitation with the hematopophyrins or derivatives for the detection of tumors[22-24]. PHOTOFRIN (dihematoprophyrin ether, DHE) is the first drug approved for human use by the FDA because of its low toxicity, under an Investigatory New Drug Exemption [24]. The data from Richter, et. al., where radiolabeled BPD-MA has been used for the same purpose, indicates that BPD-MA localizes in tumors more effectively than in other tissues, with the exception of the liver and spleen[25]. The amount of the drug retained in the liver and other organs decreases markedly and quickly with passage of time. Results of Vari et. al.[21] and those of Richter, et. al.[25] are in conformance. In results with laser induced fluorescence, spectra from the spleen displayed very low intensity. The reason for this could be a result of the high absorptive characteristics present in blood[21]. The new compounds methypheophorbides (MePH)[26], phtalocyanine (PC), and naphthalocyanin (NAPC)[27] derivatives show low phototoxicity and strong absorption at longer wavelengths. For diagnostic purposes, drugs with longer wavelength absorption (> 630 nm) are preferred, because of their deeper tissue penetration and lower natural fluorescence when excited in this range.

References

1. Fasman GD, Clevelant OH. Handbook of Biochemistry and Molecular Biolog 3d ed., CRC Press, pp. 205-210, 1975.
2. Benson RC, Meyer RA, Zaruba ME, McKhan GM. "Cellular Autofluorescence - Is it due to Flavins?", J Histochem Cytochem, 27(1), 44-48, 1979.
3. Yuanlong Y, Yanning Y, Fuming L, Yufen L, and Paozhen M. "Characteristic Autofluorescence for Cancer Diagnosis and its Origin", Lasers Surg Med, 7, 528-32, 1987.

4, 7. Schomacker KT, Frisoli JK, Compton CC, Flotte TJ, Richter JM, Nishioka NS, and Deutsch TF. "Ultraviolet Laser-Induced Fluorescence of Colonic Tissue: Basic Biology and Diagnostic Potential", Lasers Surg Med, 12, 63-78,, 1992.

5. Andersson-Engels S, Johansson J, Svanberg K, and Svanberg S. "Fluorescence Imaging and Point Measurements of Tissue: Applications to the Demarcation of Malignant Tumors and Atherosclerotic Lesions from Normal Tissue" Biochim. Photobiol., 53(6), 807-814, 1991.
6. Rava RP, Richard-Kortum R, Fitzmaurice M, Cothren R, Petras R, Sivak M, Levin H, and Feld MS. "Early detection of dysplasia in colon and bladder tissue using laser induced fluorescence" , Proc. of Optical Methods for Tumor Treatment and Early Diagnosis Mechanisms and Techniques, SPIE 1426, 68-78, 1991.
8. Pan FS, Chen S, Mintzer R, Chen CT, and Schumacker P."Studies of yeast cell oxygenation and energetics by laser fluorometry of reduced

nicotinamide adenine dinucleotide", Proc of Optical Engineerings, SPIE 1396, 5-8, 1990.

9. Schwartz JP, and Passonneau JV. "The Effect of Growth Conditions on NAD^+ and NADH Concentrations and the NAD^+ : NADH Ratio in Normal and Transformed Fibroblasts", J Biol Chem., 249(13), 4138-4143, 1974.

10. Welsh FA, O'Connor MJ, Langfitt TW. "Regions of Cerebral Ischemia Located by Pyridine Nucleotide Fluorescence", Science, 198, 951-53, 1977.

11. Beuthan J, Minet O, MÊller. "Observations of the fluorescence response of the coenzyme NADH in biological samples", Optics Letters 18(13) 1098, 1993

12. Mang TS, McGinnis C, Crean DH, Khan S, Liebow C. "Fluorescence Detection of Tumors: Studies on the Early Diagnosis of Microscopic Lesions in Pre-clinical and Clinical Studies", Proc. of Optical Methods of Tumor Treatment and Early Diagnosis; Mechanisms and Techniques, SPIE 1426, 97-110, 1991.

13, 16 Andersson PS, Montan S, Svanberg S. "Multispectral System for Medical Fluorescence Imaging" IEEE JQE, 10, 1798-1805, 1987.

14. Alfano RR, Tata DB, Cordero J, Tomashefsky P, Longo FW, and Alfano MA."Laser Induced Fluorescence Spectroscopy from Native Cancerous and Normal Tissue",IEEE JQE, 20(12), 1507-1511, 1984.

15. Tang GC, Pradhan A, and Alfano RR. "Spectroscopic differences between human cancer and normal lung and brest tissues", Lasers in Surg Med., 9, 290-295, 1989.

17. Alfano RR, Pradhan A,and Tang GC. "Optical spectroscopic diagnosis of cancer and normal breast tissue", Opt.Soc. Am. B. 6(5),1015-1023, 1989.

18. Mang T, McGinnis C, Potter W.R. " Fiberoptic Fluorescence Detection of Low Level Porphyrin Concentrations in Occult Metastases of Lymph Nodes", Proc. of Optical Fibers in Medicine IV, SPIE 1067, 289-298, 1989.

19. Andersson-Engels S, Johansson J, Svanberg S, and Svanberg K. "Fluorescence Diagnosis and Photochemical Treatment of Diseased Tissue Using Lasers: Part I" Analytical Chemistry, 61(12),1367A-1379A, 1989.

20. Andersson-Engels S, Johansson J, Svanberg S, and Svanberg K. "Fluorescence Diagnosis and Photochemical Treatment of Diseased Tissue Using Lasers: Part II" Analytical Chemistry, 62(1),19A-27A, 1990.

21. Vari SG, Papazoglou TG, Papaioannou T, Stavridi M, Pergadia VR, Fishbein MC, van der Veen MJ, Tomas R, and Grundfest WS. "Biodistribution detection of PHOTOFRIN Porfimer Sodium and Benzoporphyrin Derivative using fiberoptic sensor and ratio fluorometry" Proc of Photodynamic Therapy, SPIE 2078, 17, 1993.

22. Policard A. "Etudes sur les aspects offerts par des tumeurs expérimentales examinées a la lumiere de Wood", Jurn. Pharm. Belge, 861, 1924.

23. Lipson RL, Baldes EJ, and Olsen AM. "The Use of a Derivative of Hematoporphyrin in Tumor Detection", J Nat Cancer Inst, 26(1), 1-11, 1961.

24. Dougherty TJ. "The Future of Photodynamic Therapy: Is There One?" SPIE IS6, 1, 1990.
25. Richter AM, Cerruti-Sola S, Sternberg ED, Dolphyn D, and Levy JG. "Biodistribution of tritiated Benzoporphyrin derivative (^{3}H-BPD-MA), a new potential photosensitizer, in normal and tumor bearing mice." J Photochem Photobiol, 5, 231-244 , 1990.
26. Rogers MAJ. "Photosensitization with Deep Red Light", SPIE IS, 127, 1990.
27. Van Lier JE. "Prospect of Phtalocyanines as Photosensitizers for the PDT of Cancer", SPIE IS6, 107, 1990

Table: Summary of Excitation-Emission Maxima in Selected Biologically Important Molecules

Chromophores	**1mM Solution Dry Powder**	**(λ_{exc},λ_{em}) Maxima***
Tryptophan	**Solution**	**(275, 350 nm)**
NADH	**Solution**	**(350, 460 nm)**
NADPH	**Solution**	**(350, 460 nm)**
4-Pyridoxic Acid	**Solution**	**(300, 435 nm)**
Pyridoxal 5'-phosphate	**Solution**	**(305, 375 nm)** **(410,520 nm)**
Collagen I	**Powder**	**(340, 395 nm)** **(270, 395 nm)** **(285, 310 nm)**
Collagen III	**Powder**	**(275, 310 nm)** **(330, 390 nm)** **(370, 450 nm)**
Elastin	**Powder**	**(460, 520 nm)** **(360, 410 nm)** **(425, 490 nm)** **(260, 410 nm)**

Profilometry

3-D Shape Measurement Techniques and Their Applications

Xian-Yu Su, Jie-Lin Li
Opto-Electronics Dept., Sichuan University, Chengdu 610064, China
Phone: (+86)28 5583875-62750; Fax (+86)28 5212729;
Email:sichua@rose.cnc.ac.cn

Abstract:

This paper gives an overview of 3-D shape measurement techniques, especially those developed in our laboratory, and their applications. These techniques include moiré topography(MT), Fourier transform profilometry (FTP), phase measuring profilometry (PMP), and 360^0 shape measurement.

1. Introduction

Optical 3-D shape measurement, based on structured illumination, is becoming an interesting tool for surface topography, biomedicine, archaeology, cultural heritage documentation, machine vision, etc. Several methods have been proposed, including laser triangulation[1-6], Phase Measuring Profilometry (PMP)[7-12], Moiré Topography(MT) [13-15], Fourier Transform Profilometry (FTP) [16-18], and 360^0 shape measurement[19-22] .

This paper gives an overview of these 3-D shape measurement techniques, especially those developed in our laboratory, and describes some applications, including biomedical applications, cultural heritage documentation, and industrial applications.

2. 3-D shape measurement techniques

2.1 Laser triangulation[1-6]

In laser triangulation, including laser spot projection[1,2] and laser sheet projection[3,4], a laser spot or laser sheet is projected on the tested object, the spot (or sheet) is imaged from another direction, and the height of the illuminated spot (sheet) can be determined by the position of the spot (sheet) on the image using triangulation or a neural network[21], see Fig. 1. (a). Because the structure of laser triangulation is simple and the precision is high, it can be used widely in machine vision, 360° shape measurement, etc.. Fig. 1.(b) is the 3-D profile of a shoe last. But because laser speckle influences the depth resolution heavily, some methods have been applied to reduce speckle noise[2,5,6]. Still more effort is needed to improve the depth resolution. Another disadvantage of this method is the low speed compared with MT, PMP, and FTP.

***This project was supported by the National Natural Science Foundation of China.**

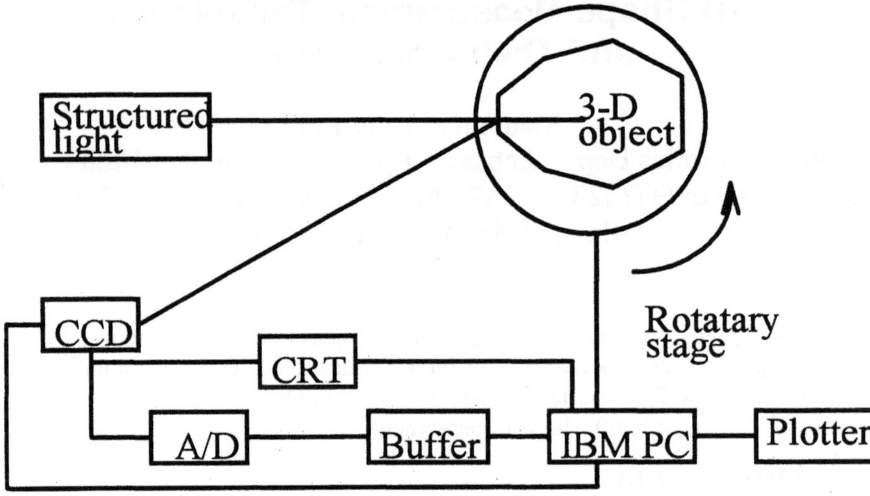

Fig.1.(a) Geometry of laser triangulation.

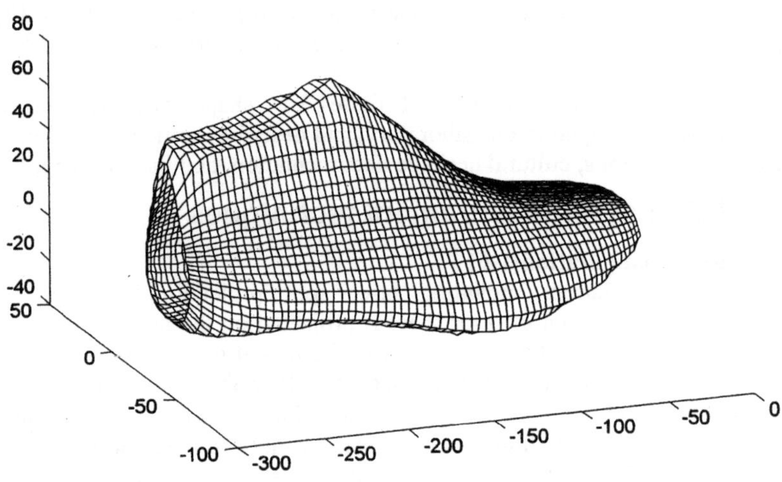

Fig.1.(b) test of a shoe last.

2.2 Phase measuring profilometry[7-12]:

In 3-D sensing techniques with structured grating illumination, fringe patterns are recorded as intensity variations. For spatial carrier fringe patterns, the intensity at any point can be written as

$$I = I_{bias}(x,y) + I_{\mathrm{mod}}(x,y)\cos[2\pi f_0 x + \phi(x,y)] \tag{1}$$

where I_{bias} is the background intensity, I_{mod} is the modulation intensity, and f_0 is the carrier frequency in the x direction. The $\phi(x,y)$ is the required phase function which characterizes the inspected shapes. For temporal and spatial carrier fringe patterns N

(N>=3) frames of patterns should be recorded. The intensity distribution in the nth phase shifted fringe pattern is given by the expression

$$I_n = I_{bias}(x,y) + I_{mod}(x,y)\cos[2\pi f_0 x + \phi(x,y) + 2\pi n / N] \qquad (2)$$

The phase function $\phi(x,y)$, calculated by any phase algorithm, is wrapped to the range $-\pi$ to π. To obtain a whole-field quantitative analysis, phase unwrapping has to be done.

The PMP method permits acquisition and processing of large amounts of data with high accuracy. This is because PMP acquires both spatial information of fringes in full field and temporal information of fringe shifting within a period of a fringe. The phase of each point is computed directly in the computer, and the height information is encoded in the phase. Using a phase-to-height table, the height can be computed[8]. Like other phase measuring techniques, a main obstacle for its application in complex object tests is improper phase unwrapping, see Fig. 2. (a), (b). Many methods have been presented in this field[10].

Intensity modulation can be regarded as the reliability of the phase unwrapping, as pointed out in the reference[9] . A higher modulation represents a higher reliability, while a lower modulation is often accompanied by local shadows, height discontinuities, and a darker background. A binary mask can be used to avoid unwrapping in the low modulation areas. If the binary mask does not cover invalid areas completely, the errors in phase unwrapping will be propagated along the path of phase unwrapping. It will introduce a false reconstruction. See Fig. 2. (c), (d).

Recently, a new phase unwrapping algorithm based on reliability analysis of data has been proposed and experimentally verified[11]. The path of phase unwrapping is determined based on the ordering of the intensity modulation in the pixels, which are on an outer layer of the boundary of the phase unwrapped areas. It means that the pixel with higher modulation will be phase unwrapped earlier. Other parameters, such as the average phase difference between neighbouring pixels, the signal-to-noise ratio, one or more band elimination filters centered around a defect region assigned by the user, could be selected as the criteria for ordering the pixels. Experimental results show that this method can be applied for complex fringe patterns with local shadows and discontinuities. Even in a worst case scenario, the errors would always be limited to local minimum areas, see Fig. 2. (e), (f).

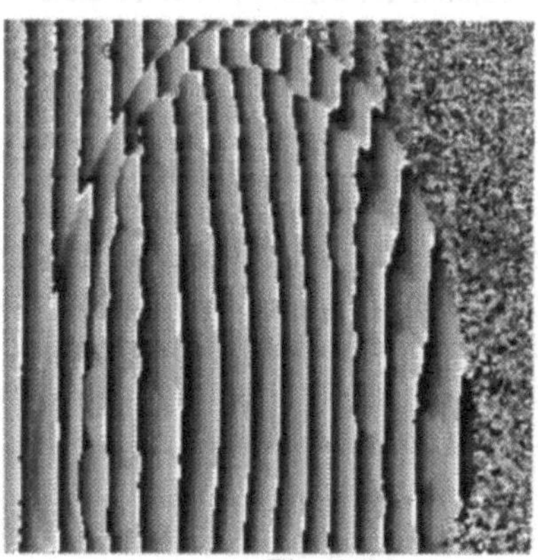

Fig. 2. (a)

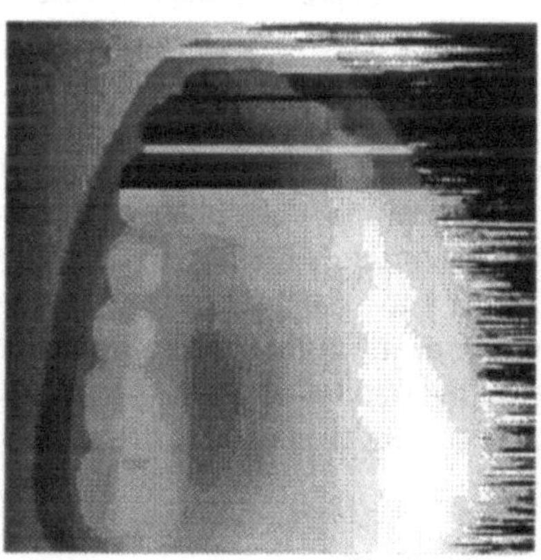

Fig. 2. (b)

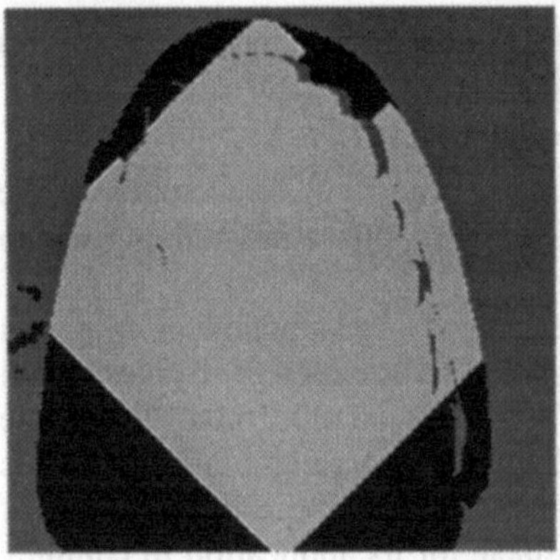

Fig. 2. (c)

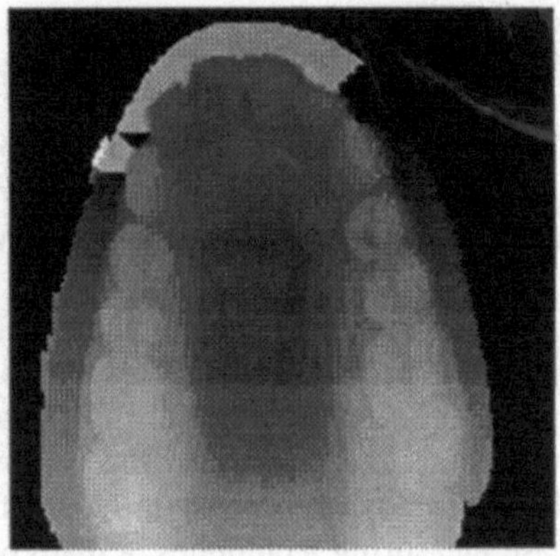

Fig. 2. (d)

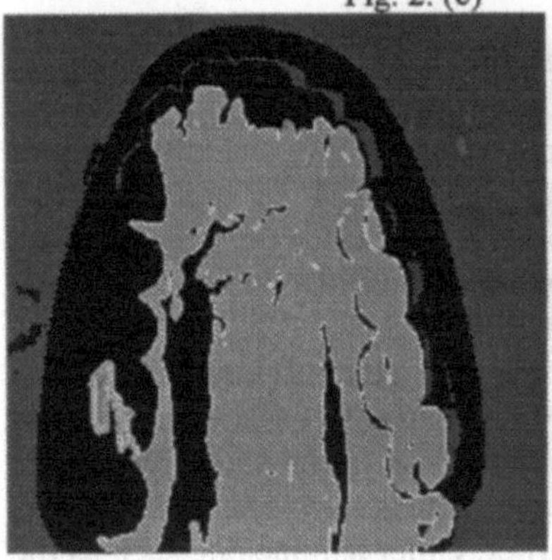

Fig. 2. (e)

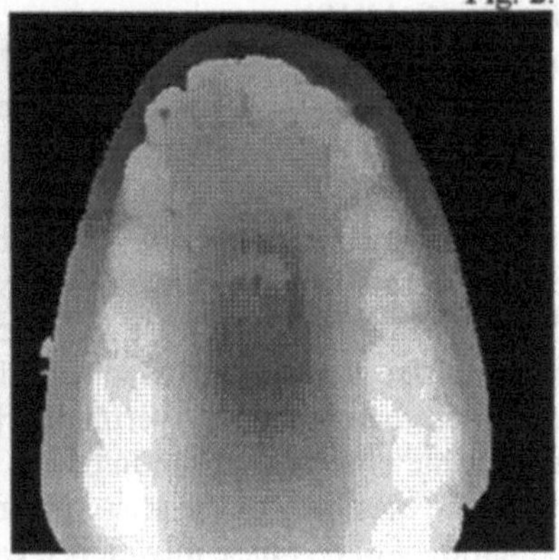

Fig. 2. (f)

Fig. 2.
(a) wrapped phase of a tooth plaster cast;
(b) unwrapping by line scan;
(c) phase unwrapping using modulation binary mask, an intermediate step;
(d) the height computed from (c); the upper bright area is wrong
(e) phase unwrapping based on modulation reliability, an intermediate step;
(f) the height computed from (e), the unwrapping is right.

In recent research[12], we have proposed a two-frequency grating projection with a liquid crystal projector. The phase of the two frequencies are computed separately. The coarse fringe pattern is used as a reference of phase unwrapping, while the thin fringe pattern is used to compute the height. In Fig. 3. (a), a hole is in the center of the platform, the depth of the hole is larger than a equivalent wavelength and can not be unwrapped by traditional methods automatically. Using the two-frequency grating method, we can get the height distribution correctly, see Fig. 3. (b).

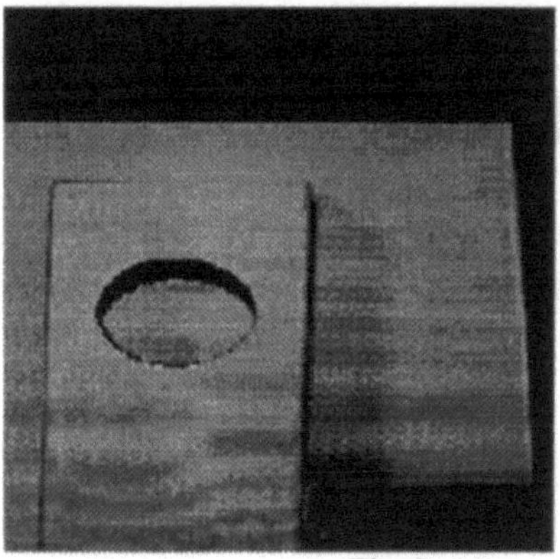

Fig. 3. (a)

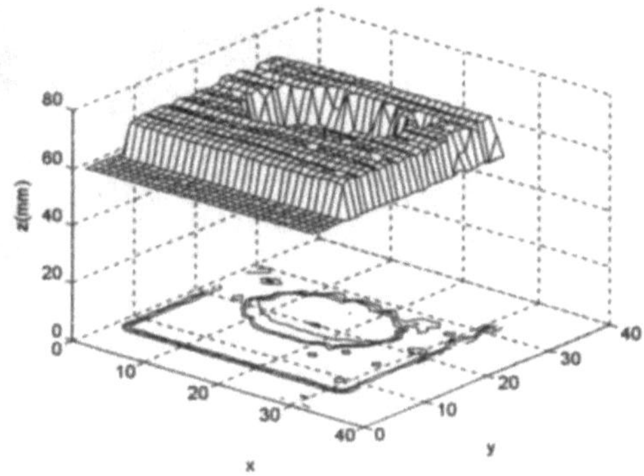

Fig. 3. (b)

Fig. 3.
(a) a platform with an isolated hole in the center; the hole is so deep that more than one fringe is shifted, it can not be unwrapped using traditional methods automatically.
(b) height computed by two frequency grating PMP;

2.3 Moiré topography[13-15]:

Moiré techniques, including in-plane moiré, projection moiré, shadow moiré, etc. use some form of structured light, typically a series of straight lines in a grating pattern, which is projected onto an object in the scene. This pattern on the object is then viewed from another angle through a secondary grating, presenting a view of the first grating line which has been distorted by the contour of the part. To determine the 3-D contour of the object, extensive software analysis and rigorous hardware manipulation to produce different moire patterns of the same object are needed. But the lateral resolution cannot exceed the spacing of the grating pattern.

2.4 Fourier transform profilometry[16-18]:

In FTP[16,17], a Ronchi grating pattern projected onto the object surface is modulated by the height distribution, and then the deformed fringe pattern is Fourier transformed and processed in its spatial frequency domain as well as in its space signal domain to demodulate the object shape from the fundamental frequency component in the Fourier spectra. FTP can fully accomplish automatic distinction between a depression and an elevation in the object shape. It requires no fringe-order assignment or fringe center determination.

A grating π phase shifting technique[17] can extend the measurable slope of height variation to nearly three times that of the unimproved FTP. Two-dimensional Fourier Profilometry(2-D FTP) [18] can provide a better separation of the height information from noise when speckle-like structures and discontinuities exist in the fringe pattern, see Fig. 4.

FTP requires only one sampling of the deformed fringe pattern, which is important for real-time data acquisition.

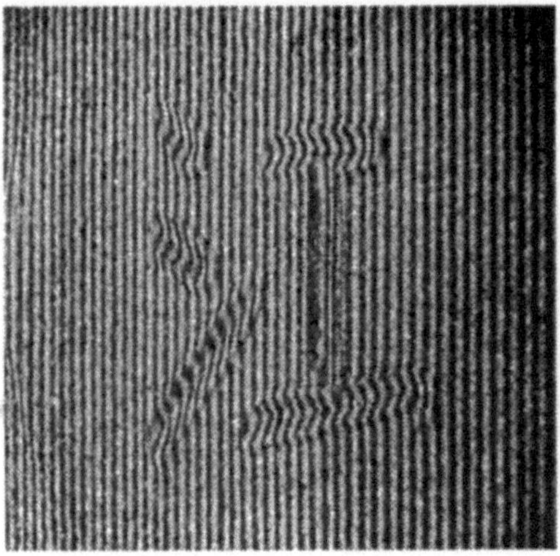

Fig. 4. (a)

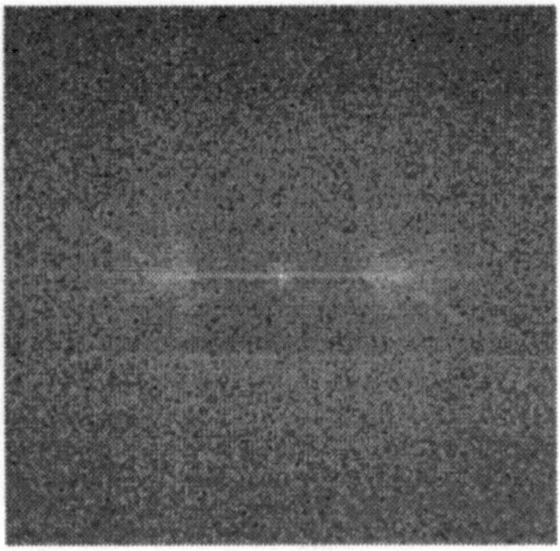

Fig. 4. (b)

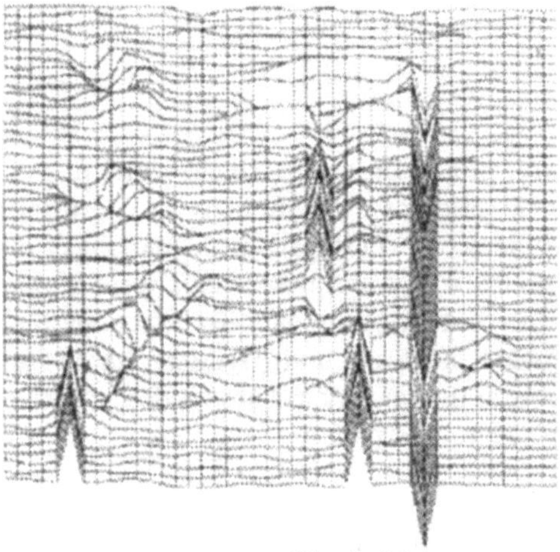

Fig. 4. (c)

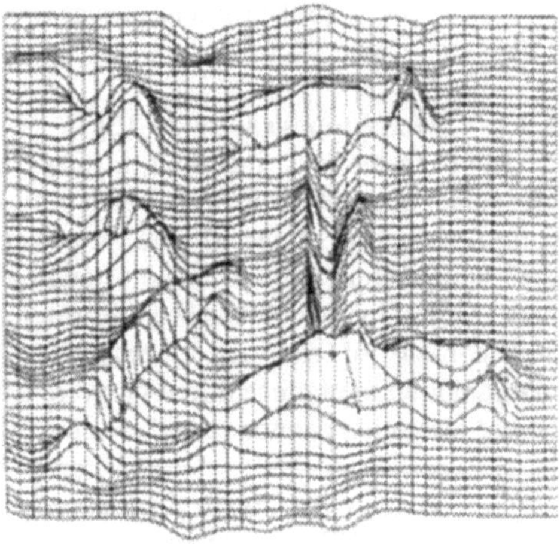

Fig. 4. (d)

Fig. 4.
(a) the surface of a Chinese character written on the sand .
(b) the 2-D Fourier spectra of (a);
(c) the restored surface of (a) using 1-D FTP;
(d) the restored surface of (a) using 2-D FTP;

2.5 360^0 shape measurement[19-22]

Most 3-D profilometries are limited to the inspection of flat objects. Complete 360^0 inspection techniques are still in the early stages of development. Generally, laser triangulation is used. The object is rotated by a rotation stage to get 360^0 profile[19]. Using some special cameras, such as time delay and integration(TDI) mode CCD, the laser sheets of different rotation angles can be recorded in a single image to form a grating pattern. Equal depth contours are derived from the deformed grating pattern by using logical moiré and scanning moiré techniques[20]. FTP[22], see Fig. 5.

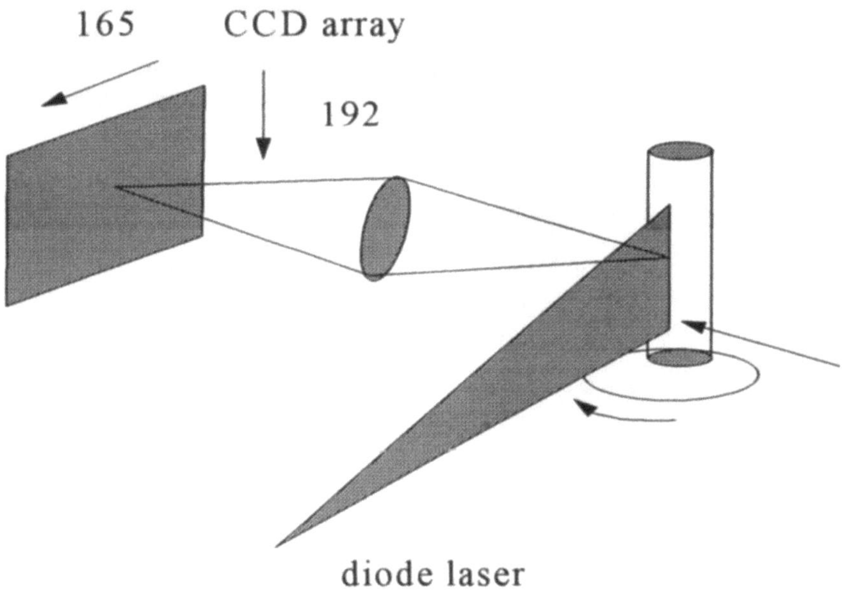

Fig. 5 (a)

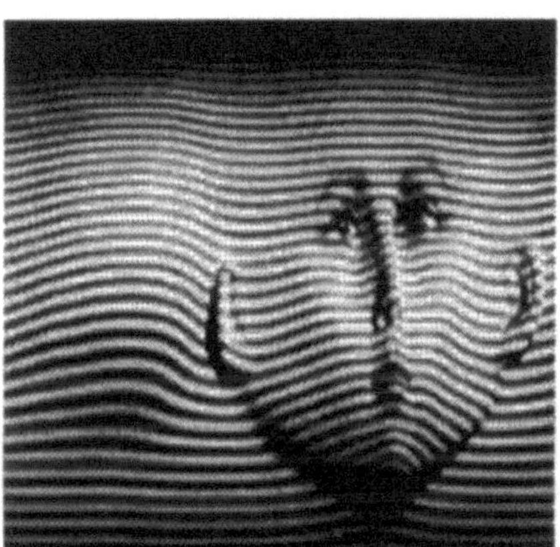

Fig. 5. (b)

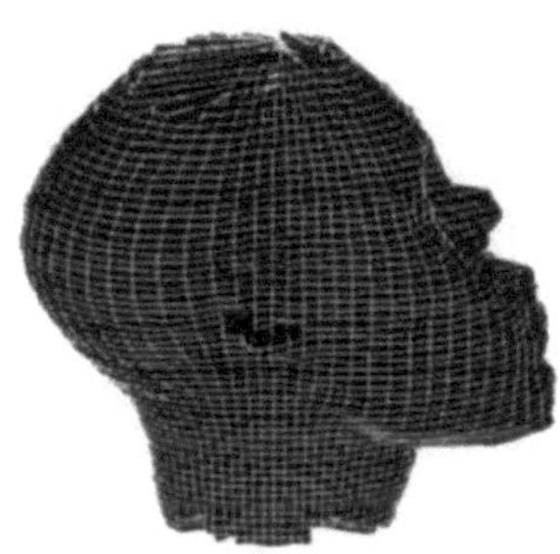

Fig. 5. (c)

Fig. 5.
(a) Geometry of a TDI camera based 360° shape measurement.
(b) The image obtained by a TDI camera.
(c) Reconstructed profile of (b) using FTP.

3.4 Others

Similar to bitmap images, a 3-D bitmap of a object may be stored in computer and transferred onto the information superhighway[23]. The 3-D data can also be edited and reconstructed by geometrical modeling techniques.

Since the 3-D data is immune to rotation and translation, it can also be used in computer vision and pattern recognition[25].

4. Conclusion

This paper gives an overview of 3-D shape measurement techniques and their applications, especially those developed in our laboratory. The essential 3-D measurement techniques are presented. Applications covering a wide spectrum including biomedical engineering, cultural heritage documentation, etc. are discussed.

This project was supported by the National Natural Science Foundation of China.

References

1. M. Rioux, G. Bechthold, D. Taylor, and M. Duggan, Opt. Eng. Vol. 26, 1245-1250 (1987)
2. W. Dremel, G. Hausler, M. Maul, Proc. SPIE. Vol. 665, 182-187(1986).
3. Gerd Hausler, Werner Heckel, Appl. Opt., Vol. 27, No. 24, (1988)
4. X.-Y. Su, B. Jia, Proc. SPIE, Vol. 1319, 362(1990).
5 R. Baribeau, M. Rioux, Appl. Opt., Vol. 30, 2873(1988).
6. J.-L. LI, X.-Y. SU, W.-S. Zhou, Optical Review, Vol. 2, 144(1995).
7. V. Srinivasan, H. C. Liu and M. Halioua, Appl. Opt. 23, 3105(1984)
8. W.-S. Zhou, X.-Y. Su, Journal of Modern Optics, Vol. 41, 89(1994)
9. X.-Y. Su, G. von Bally, D. Vukicevic, Opt. Comm., Vol. 98, 141(1993)
10. T. R. Judge, P. J. Bryanstor-Cross, Opt. and Las. in Eng., Vol. 21, 199(1994)
11. X.-Y. SU, W.-S. Zhou, Proc. SPIE, Vol. 2132, 484(1994).
12. J.-L. LI, H.-J. SU, X.-Y. SU, Appl. Opt., to be published.
13. H. Takasaki, Appl. Optics Vol. 9, 1467, (1970).
14. C. A. Sciamarella, Exp. Mech. Vol. 22, 418-433(1982).
15. A. Asundi, K. H. Yung, J. Opt. Soc. Am., Vol 8, 1591-1600(1991).
16. M. Takeda and K. Motoh, Applied Optics, Vol 24, 3977, (1983).
17. J. Li, X. Y. Su and L. R. Guo, Opt. Eng. Vol. 29, 1439,(1990).
18. J.-F. Lin, X. Y. SU, Opt. Eng. Vol. 34, 3297, (1995).
19. X-X Cheng, X.-Y. Su and L.R. Guo, Appl. Opt., Vol. 30, 1274, (1991).
20. A. Asundi and M. R. Saja, Opt. and Las. in Engi., Vol. 22, 227, (1995).
21. Ming Chang and Wen-Chih Tai, Opt. Eng., Vol. 34, No. 12, 3572, (1995).
22. X.-Y. SU, M. R. Saja, A. Asundi, Proc. SPIE, to be published (1996).
23. M. Rioux, Proc. SPIE, Vol. 2350, 2(1994).
24. W. Y. SU, X.-Y. SU, Proc. SPIE, Vol 1332,820(1990).
25. D. H. Ballard, C. M. Brown, Computer Vision, (New Jersey, 1982)

Cuneiform Surface Reconstruction by Optical Profilometry

D. Dirksen, Y. Kozlov, G. von Bally

Laboratory of Biophysics, Institute of Experimental Audiology, University of Münster, Robert-Koch-Str. 45, D-48129 Münster, Germany
E-mail: dirksen@gabor.uni-muenster.de

1 Introduction

Abstract. Computer controlled optical scanning of archaeological samples for non-contact 3D surface digitizing and investigation is discussed.

Phase-measuring profilometry is an optical technique used to acquire 3D coordinates of objects by projecting a pattern of light fringes onto their surface and analyzing the observed deformation [1], [2]. It allows non-contact scanning of samples with complex and delicate surface structures as are found e.g. with cuneiforms. The digitized data can then be processed and visualized with standard personal computer equipment to allow quantitative and qualitative investigations. With modern high capacity storage media like magneto-optical disks, the large amounts of data can be efficiently archived. Furthermore, electronic networks offer an easy method of distribution.

2 Experimental Methods

The topometric sensor consists of a projecting device and a CCD-camera, both connected to a personal computer (Fig. 1). A slide made of a metal coated glass substrate is used as a transmission grating in order to project quasi-sinusoidal fringes onto the object under investigation. Four grayscale images are captured by the CCD-camera, where each time a so called *phase shift* [3] has beeen introduced by moving the grating by a definite fraction of its spatial period. The digitized images are then used to calculate the underlying phase distribution of the fringe pattern, which in turn contains the information about its deformation by the object surface profile. From this phase distribution in turn, the height value of each picture element (pixel) is calculated by taking into account the optical and geometrical properties of the setup which have been precisely determined in advance by a calibration procedure. The achieved 3D coordinates initially form an unordered set of points (a *point cloud*) that has to be processed by a surface reconstructing algorithm before the result can be displayed, illuminated by virtual light sources, as a 3D object.

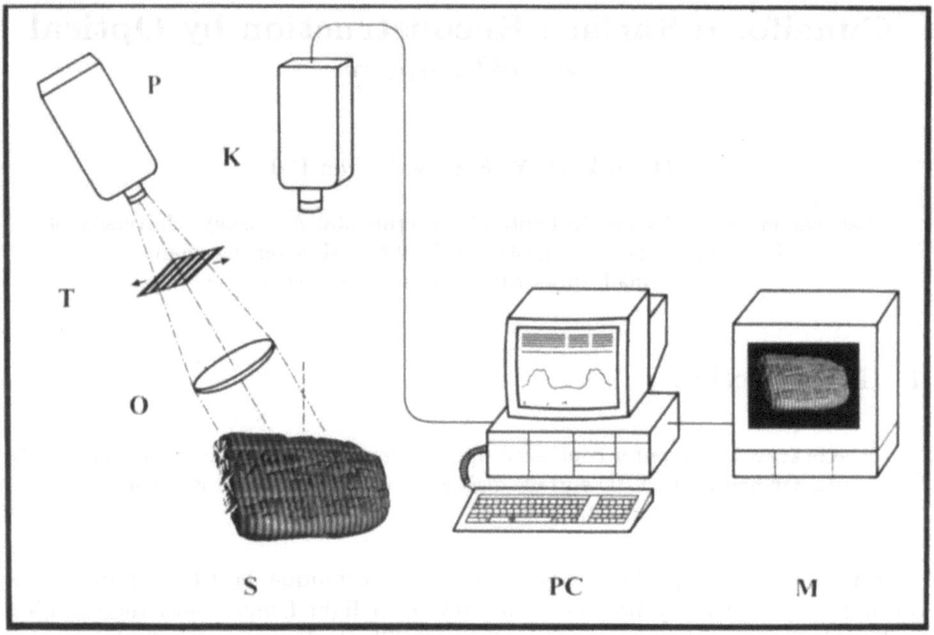

Fig. 1. Experimental Setup

3 Results

To demonstrate the application of profilometry to archaeological objects with small surface details, a cuneiform sample has been investigated. The size of the scanned area was appr. $60 \times 40\ \mathrm{mm}^2$. An achieved height resolution of $50\ \mu m$ and a lateral resolution of $100\ \mu m$ is estimated. The resulting 3D point clouds can be displayed as rendered surfaces illuminated by virtual light sources (Fig. 2). By choosing an appropriate projection, 2-dimensional height distributions (Fig. 3) can be computed and profiles along arbitrary sections may be determined (Fig. 4).

4 Discussion

Topometric 3D coordinate sensors offer — in combination with advanced computer display techniques — new possibilities for investigation as well as documentation of archaeological samples. Depending on the field size, lateral and height resolutions down to a few microns are possible. This allows a detailed reconstruction of even small surface structures as are found with e.g. cuneiform samples. The computer reconstructed surface may then be illuminated from arbitrary directions with virtual light sources to achieve an optimum 3D effect or to enhance features. Furthermore, digital data are easy to copy and to distribute.

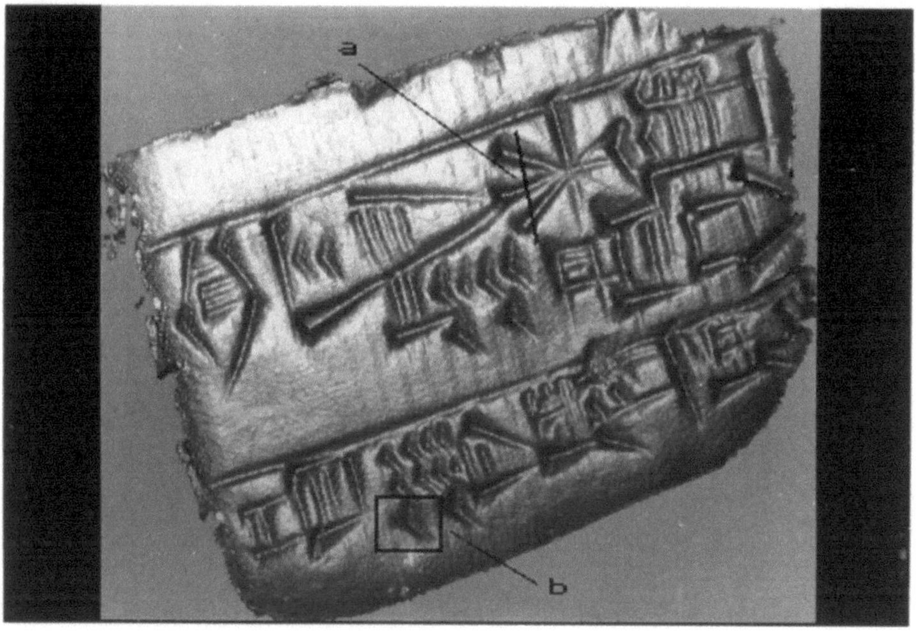

Fig. 2. Digitally reconstructed cuneiform surface

References

1. Srinivasan V., Liu H. C., Halioua M.: Automated phase measuring profilometry of 3-D diffuse objects. Appl. Opt. **23** (1984) 3105
2. Inokuchi S., Sato K.: Three-dimensional surface measurement by space encoding range imaging. Journal of Robotic Systems **2** (1985) 27
3. Creath K.: Phase measurement interferometry techniques. Prog. Opt. **26** (1988) 349

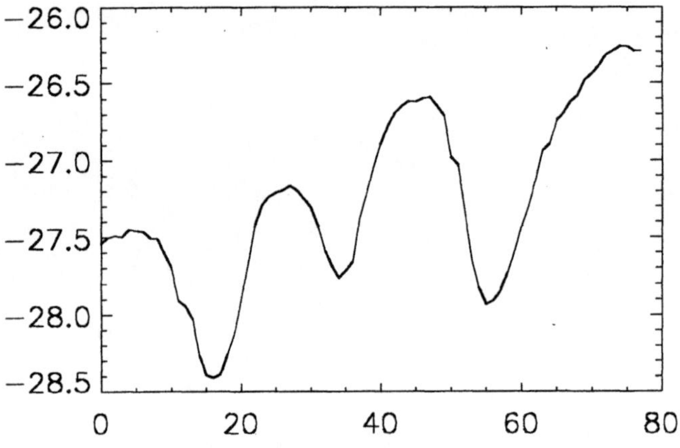

Fig. 3. Height profile along section marked (a) in Fig. 2

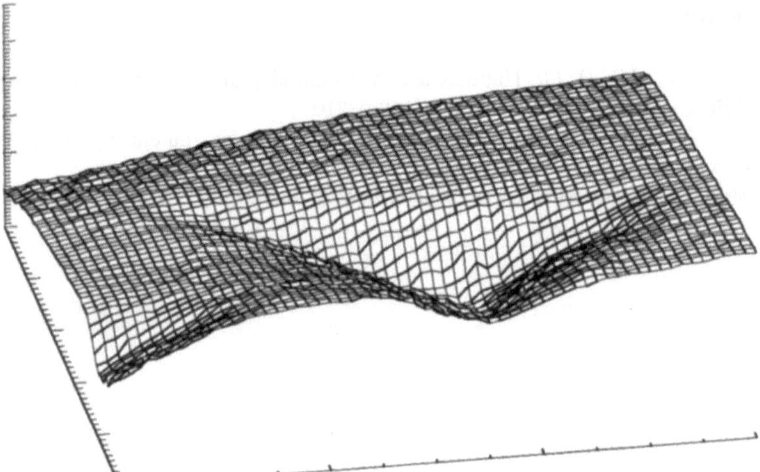

Fig. 4. Detail of height distribution of area marked (b) in Fig. 2

The Ageing of Leather: Effects on the Surface Topography

Stefan Wüstenbecker, Thomas Rose
Fachhochschule Münster, Dep. of Physical Technology
Stegerwaldstraße 39, 48565 Steinfurt, Germany

Richard Moroz
Nikolaus Kopernikus University
Sienkiewicza 30/32, 87100 Toruñ, Poland

Abstract. Changes in the topography of artificially aged leathers, e.g. used for bookbinding, were studied by analysing the Fourier-transformation of the surface profiles. First criteria relating the ageing process to significant changes in the spectra are presented.

1 Introduction

The surface topography of leather may significantly change with the ageing status of the material. The ageing status is not only influenced by the age itself, it is usually dominated by the environmental influences the leather has been exposed to during its existence. The ageing processes affect the leather surfaces beginning with hardly visible changes of the optical appearance, continuing with cracks in the surface, growing in size and area-density, so even influencing the mechanical properties of the leather, and finally lead to a complete destruction of the leather objects.

Documentation of the present status of the objects to be conserved, as well as the evaluation of the conservatory work remain incomplete if only photographic reproduction is used for documentation purposes. The photographic information is limited to two dimensions, the three dimensional micro-structure of the surface stays invisible. The knowledge of the three dimensional surface micro-structure will improve judgement of the ageing status before and after the conservatory work. Conservators will better be able to prove the effectiveness of their work, saving historical art-objects for future generations.

But even for new objects, like calf-bound books, different materials and tanning methods should be examined to find the best suited materials and production processes to save currently produced objects from known destructive influences.

This paper focuses on the identification of the slight changes of leather surfaces, resulting from artificial ageing processes that simulate the most common environmental influences.

2 Specimen Preparation and Measuring Procedures

To provide a non-destructive method to estimate environmental effects on the ageing of leather, we examined products of different leather manufacturers (Encoded #1 to #6), using different tanning methods.

Principally, mineralic, i. e. chromium or aluminium based tannage can be distinguished from various vegetably, organically based methods. The specimens examined here represent a broad spectrum of production processes, details are listed in table 1.

Sample	Organic Tannage	Mineralic Tannage
#1	-	Al / Cr
#2	Quebracho	Cr
#3	Chestnut	Al
#4	Valonea	Cr
#5	Myrobalanen	-
#6	Valonea	-

Table 1. Tannage of the examined leather specimen.

The leathers #3 and #4 were supplied by the same manufacturer, using different methods of tannage.

From each of the provided calf-leathers a set of two specimens was taken, approximately 30 × 10 mm^2 in size, one serving as a reference specimen, the other was subjected to a series of artificial ageing processes, as listed in table 2. These processes were performed by the 'Leather Conservation Centre' at Northampton (Great Britain); the subsequent chemical analysis was carried out by the 'Central Research Laboratory for Objects of Art and Science' at Amsterdam (Netherlands).

Ageing Process	1. Oxidative	2. Hydrolytic	3. Oxidative + Hydrolytic
Pre-Processing: Dry heating at 150 °C	3 Days	3 Days	3 Days
Artificial Ageing:	11 Days	14 Days	14 Days
Temperature / °C	40	40	40
Rel. Humidity / %	70	50	50
O_3 / ppm	5	-	5
SO_2 / ppm	-	25	25
NO_X / ppm	-	25	25
Light	U.V.	U.V.	U.V.

Table 2. Sequence of the three artificial ageing processes that were subsequently applied to the leather specimen.

After these ageing processes, the surface topographies of the different leather specimen and the respective reference specimen were recorded using an Autofocus Surface-Profiling Sensor (UBA 60 by UBM). On each specimen, three different areas, each sized 2×2 mm^2, were scanned with a lateral resolution of 32 pts./mm in X- and Y-direction. The resolution in height (Z) reached by the UBM-Autofocussensor is about 10 nm.

3 Analysis Methods and Results

Due to the individual structures of biological surfaces, direct comparison of the topographies appears less promising. To describe surface topographies, usually standard roughness parameters or, for a more detailed examination, a statistical analysis of the distribution of the surface normal-vectors is used.

If there are regular or cellular structures present, as can be recognised on most leather surfaces, these structures should lead to outstanding frequencies when a fourier-transformation of the surface profiles is performed. Consequently, to find impartial criteria that correlate to the ageing status, the measured topographies have been fourier transformed using a 2-dimensional Fast-Fourier-Transformation (FFT) algorithm.

The analysis of the resulting power spectra was performed with respect to the total power, i.e. the sum of the square of all matrix coefficients (Figure 1) and the frequency centroid (Figure 2).

3.1 Total Power Analysis

We found that, except for #3 and #4 (provided by the same manufacturer), the total power of the spectra significantly increases with the ageing process. Those leathers, already initially showing high values of the total power (#1 and #2) revealed the strongest absolute increase (Figure 1). Both manufacturers use chromium tannage in combination with other mineralic or vegetable-based methods.

The pure vegetably tanned leathers (#5 and #6) show relatively low total powers, even after ageing. However, leather #5 shows the lowest total power when new, but the strongest relative increase (~ 400%) due to the artificial ageing.

Best stability in combination with relatively low total power is achieved by the leather #4 produced with a combination of vegetably and chromium based tannage.

3.2 Frequency Centroid Analysis

The behaviour of the frequency centroids does not show a uniform tendency. As can be seen from figure 2, the frequency centroids shows a relatively large variation for the new leathers, not allowing the calculation of an average centroid frequency. For the aged leathers the variation of contriod frequencies is much smaller. The centroid frequency seems to approach a constant value of 1,69 ± 0,14 1/mm, corresponding to an average structure size: 0,59 ± 0,05 mm.

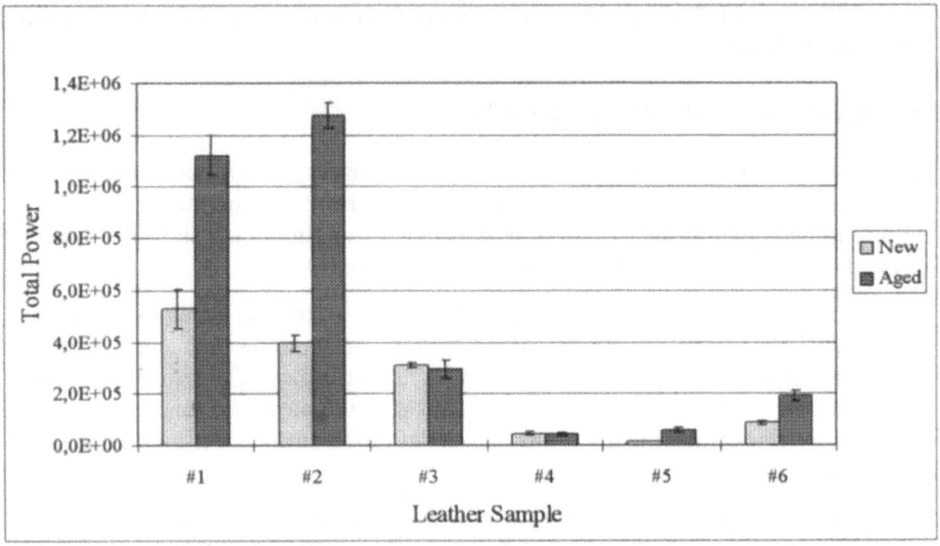

Figure 1. Analysis of the Total Power of the fourier-transformed leather surface profiles.

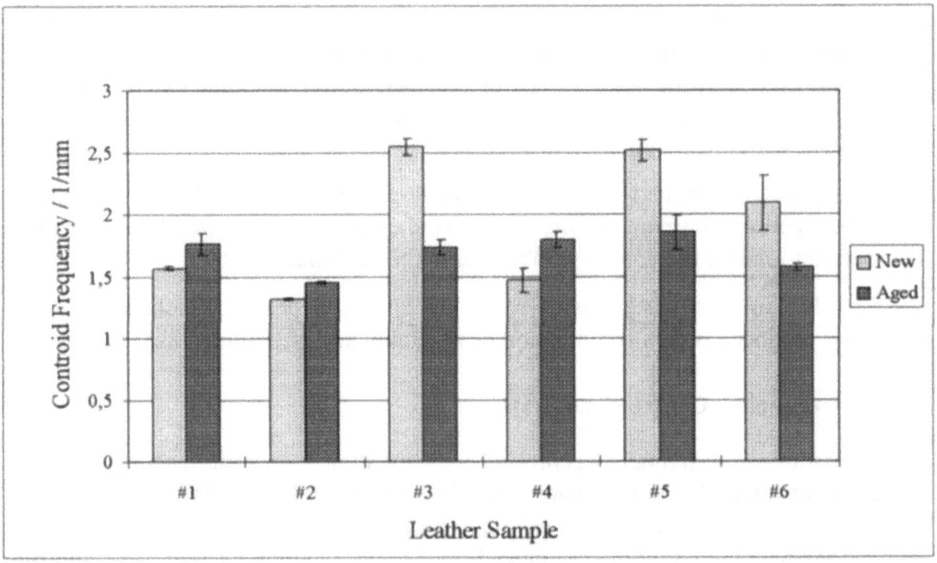

Figure 2. Analysis of the Frequency Centroids of fourier-transformed leather surface profiles.

This fact might, for ancient objects, be a significant sign to distinguish between leathers of different biological provenance.

4 Summary

It could be shown that already the very beginning of the ageing of leather, only visible as a slight change in the colour of the leather, can be detected by topographic methods.

Significant differences can be realised between different production methods. Further research will supply more detailed information on how the tannage process influences the long term stability of the leather.

The average structure size, measured as the frequency centroid of the power spectra, seems to approach a constant value with the ageing process for the specimen examined here. Additional investigation will show whether this measure can serve to identify the biological provenance of ancient leathers.

Authors' Index

MIX
Papier aus verantwortungsvollen Quellen
Paper from responsible sources
FSC® C105338

If you have any concerns about our products,
you can contact us on
ProductSafety@springernature.com

In case Publisher is established outside the EU,
the EU authorized representative is:
Springer Nature Customer Service Center GmbH
Europaplatz 3, 69115 Heidelberg, Germany

Printed by Libri Plureos GmbH
in Hamburg, Germany